High-Speed Rail

Edited by Hamid Yaghoubi

Published in London, United Kingdom

IntechOpen

Supporting open minds since 2005

High-Speed Rail
http://dx.doi.org/10.5772/intechopen.76549
Edited by Hamid Yaghoubi

Contributors
Diego Ferreño, Isidro A. Carrascal, José A. Casado, Soraya Diego, Estela Ruiz, Lu Zhou, Yi-Qing Ni, Xiao-Zhou Liu, Alexey Vereschaka, Alexey Popov, Sergey Grigoriev, Mikhail Kulikov, Catherine Sotova, Hamid Yaghoubi

Notice
Statements and opinions expressed in the chapters are these of the individual contributors and not necessarily those of the editors or publisher. No responsibility is accepted for the accuracy of information contained in the published chapters. The publisher assumes no responsibility for any damage or injury to persons or property arising out of the use of any materials, instructions, methods or ideas contained in the book.

First published in London, United Kingdom, 2019 by IntechOpen
IntechOpen is the global imprint of INTECHOPEN LIMITED, registered in England and Wales, registration number: 11086078, The Shard, 25th floor, 32 London Bridge Street
London, SE19SG – United Kingdom
Printed in Croatia

British Library Cataloguing-in-Publication Data
A catalogue record for this book is available from the British Library

Additional hard and PDF copies can be obtained from orders@intechopen.com

High-Speed Rail
Edited by Hamid Yaghoubi
p. cm.
Print ISBN 978-1-83880-922-5
Online ISBN 978-1-83880-923-2
eBook (PDF) ISBN 978-1-83880-924-9

We are IntechOpen,
the world's leading publisher of
Open Access books
Built by scientists, for scientists

4,200+
Open access books available

116,000+
International authors and editors

125M+
Downloads

Our authors are among the

151
Countries delivered to

Top 1%
most cited scientists

12.2%
Contributors from top 500 universities

Interested in publishing with us?
Contact book.department@intechopen.com

Numbers displayed above are based on latest data collected.
For more information visit www.intechopen.com

Meet the editor

Dr. Hamid Yaghoubi is the director of Iran Maglev Technology (IMT). He became Iran's top researcher in 2010. In this regard, he was awarded by the Iranian president, the Iranian Minister of Science, Research and Technology, and the Iranian Minister of Information and Communication Technology. Dr. Yaghoubi became the 2011 and 2012 Outstanding Reviewer for the *Journal of Transportation Engineering* (JTE), American Society of Civil Engineers (ASCE), USA. One of his journal papers became the 2011 Top Download Paper for *JTE*. He received the ICCTP2011 award for the 11th International Conference of Chinese Transportation Professionals (ICCTP2011), ASCE. He is an assistant chief editor and editorial board member of a number of journals; a reviewer of many journals, books, and conferences; and an editor of a number of books. Dr. Yaghoubi has cooperated with many international conferences as chairman, keynote speaker, chair of session, publication chair, and member of committees, including scientific, organizing, steering, advisory, and technical program. He is also a member of several international committees.

Contents

Preface

The rapid expansion of transportation industries worldwide, including railways, and the never-ending desire to reduce travel time have highlighted the need to resort to advanced transit systems. Conventional railway systems have been modified to make travel at much higher speeds possible. *High-Speed Rail* includes the main topics and basic principles of high-speed railways (HSR). The book reflects new engineering and track developments, the most current design methods, as well as the latest industry standards and policies. This book provides a comprehensive overview of the significant characteristics for HSR, highlighting recent advancements, requirements, and improvements, and details of the latest techniques in the global market. *High-Speed Rail* contains a collection of the latest research developments on HSR, comprehensively covering basic theory and practice in sufficient depth to provide a solid grounding for railway engineers. The book will also help readers to maximize effectiveness in all facets of HSR. It is a professional book and a credible source and valuable reference that will be applicable and useful for all professors, researchers, engineers, practicing professionals, trainee practitioners, students, and others interested in HSR. The book consists of four chapters.

Chapter 1 reviews HSR. HSR is defined as an intercity passenger transit system that is time competitive with aircraft and/or automobiles on a door-to-door basis. The fundamental reason for considering the implementation of rapid transit systems is higher speed, which can easily equate to shorter travel time. Therefore, there is a need to look at the technical specifications of each technology by examining the potential improvement of each technology in terms of speed, travel time, and other advantages. People have always demanded reductions in travel time for many good reasons, such as trade, leisure, etc. This has forced the rapid expansion of transportation industries worldwide, including railways. Consequently, high-speed transit systems have been introduced in many countries. These systems are manufactured based on advanced engineering methods and technologies. The congestion in transportation modes associated with increased travel has caused many problems, including public concerns, among which are prolonged travel time, growing accident rates, worsening environmental pollution, and accelerating energy consumption. On the contrary, high-speed ground transportation, characterized by high speed, operating reliability, passenger ride comfort, and an excellent safety record, is considered one of the most promising solutions to alleviate congestion.

Chapter 2 includes an overview of the different elements present in the railway superstructure of high-speed lines in Spain. The performance of rail transport has increased significantly in recent decades, in particular due to the gradual introduction of HSR worldwide. In 1981, the world's first high-speed line was inaugurated (Paris–Lyon, 425 km); nowadays, high-speed networks are in operation in more than 20 countries covering more than 35,000 km (with more than 25,000 additional km under construction). Spain is one of the leading countries implementing high-speed rail. In fact, Spain is the second country by total distance of railways installed (only behind China) and the first relative to population and surface. Since the installation of the first high-speed line in Spain in 1992, the elements of the superstructure (sleepers and fastening system) have undergone a continuous evolution to improve performance, ensure durability of the components, and maximize the comfort of passengers. This evolution is due to an adequate selection of materials based on the characterization of their physical and mechanical properties to ensure optimum in-service conditions. Throughout this chapter, the innovations that have been incorporated over time are analyzed, as well as the methods used to validate them. In particular, a description of the mechanical characterization procedures is presented.

Chapter 3 evaluates contemporary inspection and monitoring for high-speed rail systems. Non-destructive testing (NDT) techniques have been explored and extensively utilized to help maintain safe operation and improve ride comfort of the rail system. In line with the ascension of NDT techniques, structural health monitoring (SHM) brings a new era of real-time condition assessment to rail systems without interrupting train services, which is significant to HSR. This chapter first gives a review of the NDT techniques for wheels and rails, followed by recent applications of SHM on HSR enabled by a combination of advanced sensing technologies using optical fiber, piezoelectric, and other smart sensors for onboard and online monitoring of railway systems from vehicles to rail infrastructure. An introduction to the research frontier and development direction of SHM on HSR is also provided and concerns both sensing accuracy and efficiency through cutting-edge data-driven analysis studies embracing wireless sensing and compressive sensing, which endorse big data's role in this new age of transport.

Chapter 4 considers the methods to increase the performance and reliability of reprofile machining of wheel tread profiles. Looking beyond both milling and turning, the cutting tool is the key element to ensure performance and reliability of the manufacturing process, and the study considers methods to increase the performance properties of cutting tools. In particular, the study includes the investigation of ways to improve cutting tools (carbide inserts) to machine wheel tread profiles.

Dr. Hamid Yaghoubi
Director of Iran Maglev Technology,
Iran

Chapter 1

Introductory Chapter: High-Speed Railways (HSR)

Hamid Yaghoubi

1. High-speed railways

High-speed railways (HSR) are defined as an intercity passenger transportation system that is time-competitive with air and/or auto on a door-to-door basis. The main reason for considering the implementation of rapid transportation systems is higher speed, which can easily equate to shorter travel time. The rapid expansion of transportation industries worldwide, including railways, and the never-ending desire to reduce travel time have highlighted the need to resort to the advanced transit systems. Conventional railway systems have been modified to make them travel at much higher speeds. People have always demanded reduction in travel time for many good reasons such as trade, leisure, etc. This has forced rapid expansion of transportation industries worldwide, including railways. Consequently, high-speed transit systems have been introduced in many countries. These systems are manufactured based on advanced engineering methods and technologies. Rapid transit systems must fulfill the major elements of the transport politics. The main aims consist in the increase of speed in the transportation corridors, flexibility, environmental acceptance, ride comfort, stresses (noise, pollutions, and vibrancies), etc. Mobility and transportation infrastructure guarantee a high grade of freedom and quality for the citizens, for their work, and leisure time. Infrastructure is an important location factor in the regional and global sense. It strongly influences the development of the society and the growth of the national economies. The mobility of individuals is impossible without an equivalent volume of traffic and

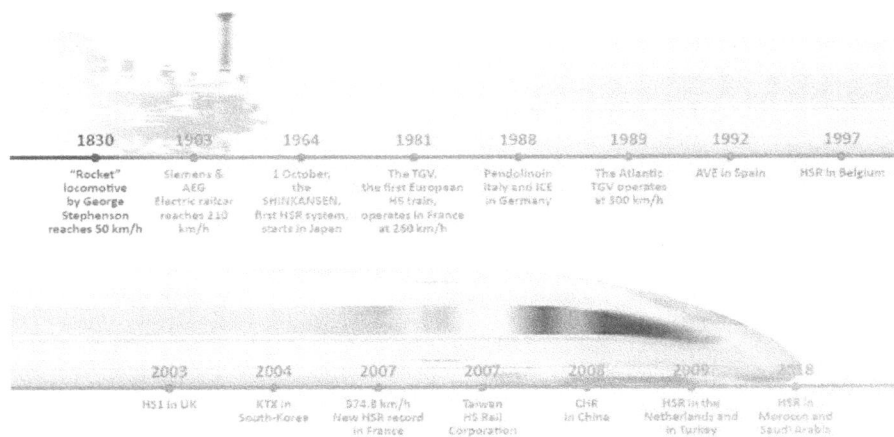

Figure 1.
History of high-speed rail (HSR).

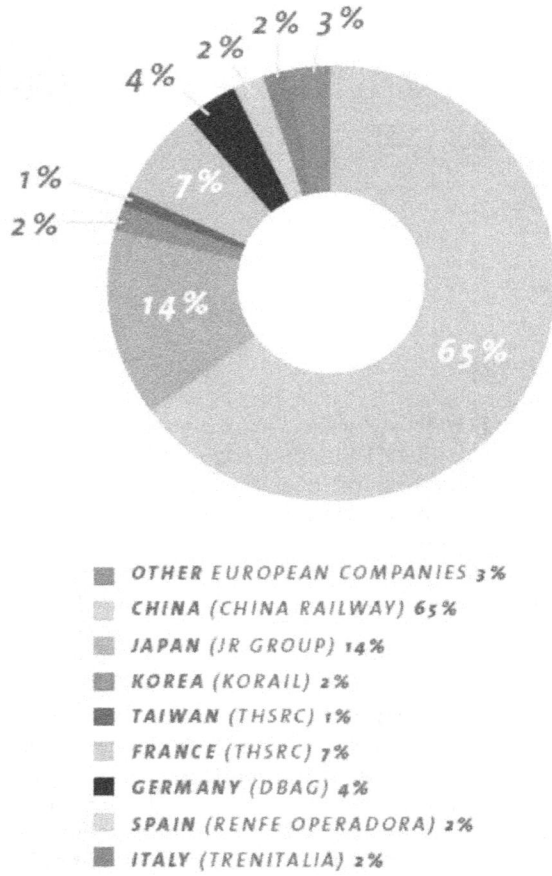

OTHER EUROPEAN COMPANIES *3%*
CHINA *(CHINA RAILWAY)* *65%*
JAPAN *(JR GROUP)* *14%*
KOREA *(KORAIL)* *2%*
TAIWAN *(THSRC)* *1%*
FRANCE *(THSRC)* *7%*
GERMANY *(DBAG)* *4%*
SPAIN *(RENFE OPERADORA)* *2%*
ITALY *(TRENITALIA)* *2%*

Figure 2.
HSR market shares in 2016 (PASSENGERS.KILOMETER).

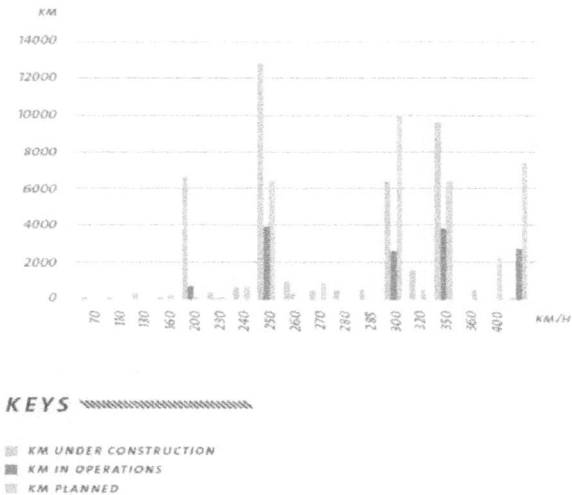

KEYS

KM UNDER CONSTRUCTION
KM IN OPERATIONS
KM PLANNED

Figure 3.
High-speed rail network.

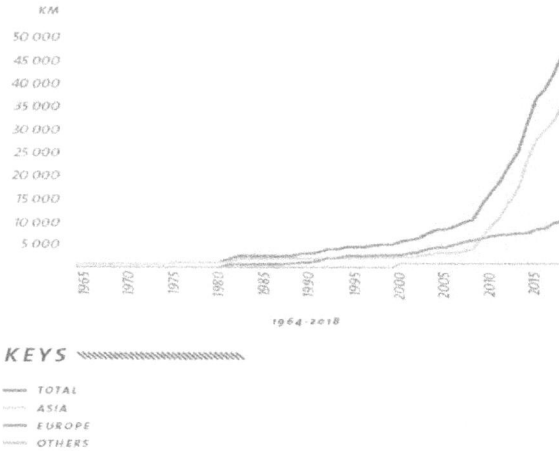

Figure 4.
High-speed rail network length.

transportation infrastructure. Urban developments lead to a considerable increase of the road and an increase of stresses for the people and environment. The public transportation policy must be faced up to this challenge and act appropriately in time. A major vision is the development of HSR, which can relocate certain parts of the road and air traffic to these systems and to enhance growth of congested urban areas and coalescence of the area. Examples of HSR include the French Train à Grand Vitesse (TGV), the Japanese Shinkansen, the German Intercity Express (ICE), the Spanish AVE, etc. [1–17] (**Figures 1–4**).

Acknowledgements

This work was performed by Iran Maglev Technology (IMT).

Author details

Hamid Yaghoubi
Iran Maglev Technology (IMT), Tehran, Iran

*Address all correspondence to: info@maglev.ir

IntechOpen

References

[1] Iran Maglev Technology (IMT), Tehran, Iran. 2008. Available from: www.maglev.ir

[2] Yaghoubi H. Magnetically Levitated Trains, Maglev. Tehran, Iran: Pooyan Farnegar Publisher; 2008. ISBN: 978-600-5085-05-1

[3] Yaghoubi H, Sadat HM. Mechanical assessment of maglev vehicle—A proposal for implementing maglev trains in Iran. In: The ASME 10th Biennial Conference on Enginveering Systems Design and Analysis (ESDA), Vol. 2. Yeditepe University, Istanbul, Turkey; 2010. pp. 299-306. ISBN: 978-0-7918-4916-3

[4] Yaghoubi H, Ziari H. Assessment of structural analysis and design principles for maglev guideway: A case-study for implementing low-speed maglev systems in Iran. In: The 1st International Conference on Railway Engineering, High-Speed Railway, Heavy Haul Railway and Urban Rail Transit. Beijing Jiaotong University, Beijing, China: China Railway Publishing House; 2010. pp. 15-23. ISBN: 978-7-113-11751-1

[5] Behbahani H, Yaghoubi H. Procedures for safety and risk assessment of maglev systems: A case-study for long-distance and high-speed maglev project in Mashhad-Tehran route. In: The 1st International Conference on Railway Engineering, High-speed Railway, Heavy Haul Railway and Urban Rail Transit. Beijing Jiaotong University, Beijing, China: China Railway Publishing House; 2010. pp. 73-83. ISBN: 978-7-113-11751-1

[6] Yaghoubi H. The most important advantages of magnetically levitated trains. Towards Sustainable Transportation Systems. In: Proceedings of the 11th International Conference of Chinese Transportation Professionals (ICCTP2011). Nanjing, China: American Society of Civil Engineers (ASCE) Publisher; 2011. pp. 3974-3986. ISBN: 978-0-7844-1186-5

[7] Yaghoubi H, Ziari H. Development of a maglev vehicle/guideway system interaction model and comparison of the guideway structural analysis with railway bridge structures. ASCE, Journal of Transportation Engineering. 2011;**137**(2):140-154

[8] Yaghoubi H, Barazi N, Kahkeshan K, Zare A, Ghazanfari H. Technical comparison of maglev and rail rapid transit systems. In: The 21st International Conference on Magnetically Levitated Systems and Linear Drives (MAGLEV 2011). Daejeon Convention Center, Daejoen, Korea; 2011

[9] Yaghoubi H, Rezvani MA. Development of maglev guideway loading model. ASCE, Journal of Transportation Engineering. 2011;**137**(3):201-213

[10] Behbahani H, Yaghoubi H, Rezvani MA. Development of technical and economical models for widespread application of magnetic levitation system in public transport. International Journal of Civil Engineering (IJCE). 2012;**10**(1):13-24

[11] Yaghoubi H, Barazi N, Aoliaei MR. Chapter 6: Maglev. Infrastructure Design, Signalling and Security in Railway. University Campus STeP Ri. Rijeka, Croatia: InTech; 2012. pp. 123-176. ISBN: 978-953-51-0448-3

[12] Yaghoubi H. Practical Applications of Magnetic Levitation Technology. Final Report. Iran Maglev Technology (IMT), Tehran, Iran. 2012. Available from: http://www.maglev.ir/eng/documents/reports/IMT_R_22.pdf

[13] Yaghoubi H, Keymanesh MR. Design and evaluation criteria for stations of magnetically levitated trains. Journal of Civil Engineering and Science (JCES), World Academic Publishing (WAP), Hong Kong. 2013;**2**(2):72-84. ISSN: 2227-4634 (Print). ISSN: 2227-4626 (online)

[14] Yaghoubi H. The most important maglev applications. Journal of Engineering, Hindawi Publishing Corporation, New York, USA. 2013;**2013**:537986. DOI: 10.1155/2013/537986. 19p. ISSN: 2314-4912 (Print). ISSN: 2314-4904(Online)

[15] Yaghoubi H. Application of magnetic levitation technology in personal transportation vehicles. Current Advances in Civil Engineering (CACE), American V-King Scientific Publishing, New York, USA. 2013;**1**(1):7-11

[16] Yaghoubi H. Urban Transport Systems. University Campus STeP Ri. Rijeka, Croatia: InTech Publisher; 2017. DOI: 10.5772/62814. ISBN: 978-953-51-2873-1 (Print). ISBN: 978-953-51-2874-8

[17] Leboeuf M. High Speed Rail, Fast Track to Sustainable Mobility. Publication UIC Passenger Department; 2018. High Speed Rail Brochure, International Union of Railways (UIC)

Chapter 2

Optimization of Components of Superstructure of High-Speed Rail: The Spanish Experience

Estela Ruiz, Isidro A. Carrascal, Diego Ferreño,
José A. Casado and Soraya Diego

Abstract

The performance of rail transport has increased significantly in recent decades, in particular due to the gradual introduction of high-speed rails worldwide. In 1981, the first high-speed line of the world was inaugurated; nowadays, high-speed is operating in more than 20 countries, the high-speed network covering more than 35,000 kms (with more than 25,000 additional kms under construction). Spain is the second country by total distance of railways installed (only behind China) and the first in terms relative to the population and surface. Since the installation of the first high-speed line in Spain in 1992, the elements of the superstructure have undergone a continuous evolution, in order to improve the performance, the durability of the components and the comfort of the passengers. This evolution rests on an adequate selection of materials based on the characterization of their physical and mechanical properties to ensure the optimum in-service conditions. This chapter includes an overview of the different elements present in the railway superstructure of the high-speed lines in Spain. Throughout the text, the innovations incorporated over time are analyzed, as well as the methods used to validate them. In particular, a description of the mechanical characterization procedures is presented.

Keywords: railway, high-speed superstructure, fastening system, mechanical properties, torque, temperature, moisture, corrosion

1. Introduction

Spain occupies the second place in the world, behind China, in kilometers of high-speed railway (HSR) built. The Spanish HSR that links the cities of Madrid and Barcelona was designed to travel at an average speed of 350 km/h in order to compete with the airplane. This implies an increase in average speed of 25% as compared to its predecessor, the railway that connects Madrid and Seville. Consequently, the dynamic forces exerted on the track and all its components have significantly increased.

One of the essential constituent elements that determine the quality of the track is the fastening system that connects the rails and the sleepers which, in turn, transmits both the static and dynamic forces exerted by the passing stocks to the railway infrastructure. Originally, the fastening elements were designed to prevent the overturning and the transverse displacement of the rails. The advances in railway technology have compelled the fastening systems to fulfill new require-ments such as the maintenance of the track gauge or the resistance to longitudinal loads due to the thermal expansion of the welded rails. In addition, the increased use of concrete sleepers has led to the need for the presence of an elastic element between the sleeper and the rail, the rail pads, to cushion the impacts and to reduce the high stiffness of the concrete sleepers.

Currently, the elastic fastening system between rails and sleepers is composed of metal spring clips, insulating plates of a polymeric nature and anchoring screws. These elements fix the rail to the sleeper minimizing the longitudinal and lateral displacements, as well as rotation, that are produced by the transversal, vertical, and longitudinal forces transmitted by the wheels of the passing vehicles. In addi-tion, the fastening system provides the elastic response needed to counteract the vertical wave movements of the track that would give rise to high dynamic forces between wheels and rails avoiding vibrations and noise. Likewise, it maintains the gauge of the track and the inclination of the rails within the admissible tolerances avoiding the overturning of the rail.

The tightening of the fastener that holds the rail to the sleeper plays an essential role in guaranteeing the performance of the whole assembly. In this sense, the grip must provide a resistance to longitudinal displacements higher than the friction between the ballast and the sleeper; in addition, it has to reduce the movements of the head of the rail with a safety coefficient. At the same time, it must withstand the stresses to which it may be subjected to under in-service or accidental conditions. Finally, the electrical insulation between the two rails on electrified lines or equipped with signaling systems must be guaranteed.

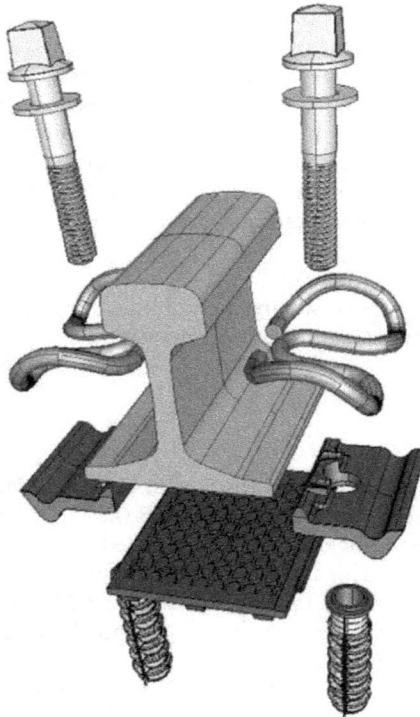

Figure 1.
Sketch of the VM fastening system [2].

Among the existing types, the VM fastening system (see **Figure 1**) is the one installed in the Madrid-Barcelona HSR. Each of the components of this system (tension clips, guide plates, rail pad, plastic dowels, and steel screws) has been optimized regarding its mechanical behavior [1]. This chapter summarizes the studies carried out at LADICIM (Laboratory of Science of Materials of the University of Cantabria) over more than two decades, to improve the mechanical response of each of the constituents of the VM fastening system, in order to optimize the response of the assembly under in-service or accidental conditions. Likewise, the influence of parameters such as the tightening torque on the mechanical response of the system was studied. Finally, the deterioration undergone by the constitutive materials because of the action of external agents (humidity and temperature as a source for corrosion of metals or degradation of polymers) was analyzed in depth.

2. Polyamide guide plates

The mechanical characterization of this type of components is based on the application of three types of loadings: static, impact, and fatigue. In order to optimize this component, an intense experimental campaign has been carried out to analyze the influence of parameters such as the moisture content of the plate, the tightening of the assembly or the mechanical aging of the plate.

For this study, A2-type guide plates injected with polyamide 6.6 reinforced (35% wt.) with short fiber glass were used, all corresponding to the same manufacturing series. The plates were received under the dry-as-molded (DAM) condition after injection and, subsequently, they were subjected to a treatment to gain different moisture contents. A first group remained in an oven at 100°C before testing. At the same time, another group of plates remained in the laboratory environment reaching a moisture content (% wt.) of 0.5%. Three sets of plates were submerged in a water bath at 50°C until reaching humidity contents of, respectively, 1.8, 2.2, and 3.6%. For the characterization of the plates, a specific device setup was developed and manufactured simulating both the support on the sleeper, the fastening system, the tightening of the clip, and the action of the base of the rail.

2.1 Static response

To carry out the tests, an increasing load was applied by means of a universal servo-hydraulic test machine with a loading capacity of ±250 kN under control of displacement at a speed of 0.05 mm/s. A linear variable displacement transducer (LVDT) was used to record the displacement between the base of the rail and the frame where the guide plate was located.

2.1.1 Influence of the moisture content on the static behavior

To study the influence of the moisture content, plates belonging to each of the groups described above were used. **Figure 2** shows the results obtained in the static test on plates with different degrees of humidity. As can be verified, except for the dry plate which breaks when reaching a certain load value, the rest of the plates present a point of inflection. This phenomenon is due to the fact that increasing the moisture content of the plate increases its deformability and, at some moment in the loading process, the inner part of the hole contacts the screw, and then, they collaborate jointly against the displacement of the rail base.

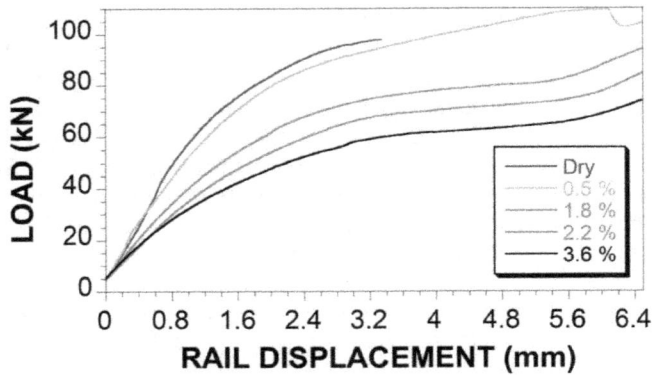

Figure 2.
Results of the static tests conducted on plates with different levels of humidity.

As can be seen in all cases, the contact of the plate with the screw is approximately constant, around 4.4 mm, although, as the plate humidity increases, the force decreases, as the plate becomes more and more flexible.

2.1.2 Influence of the tightening torque on the static behavior

To assess the influence of the tightening torque on the static behavior of the plates, a series of tests was performed applying the following three torques: 100, 250, and 350 N·m. In all cases, plates with a moisture content of 1.8% were employed. The influence of the tightening torque on the static behavior of the plates is represented in **Figure 3**.

As can be appreciated, the curves have the same appearance in the three cases; all the curves show the inflection point that represents the contact point between the plate and the screw. Again, this inflection point occurs for a displacement of 4.4 mm.

On the other hand, the values of the force at the contact point decrease with the tightening torque: 71.4, 79.0, and 80.6 KN for 100, 250, and 350 N·m, respectively. Since the plates have the same moisture content, the fact that the lower the tightening torque, the easier the contact, can be explained because the plate can slide

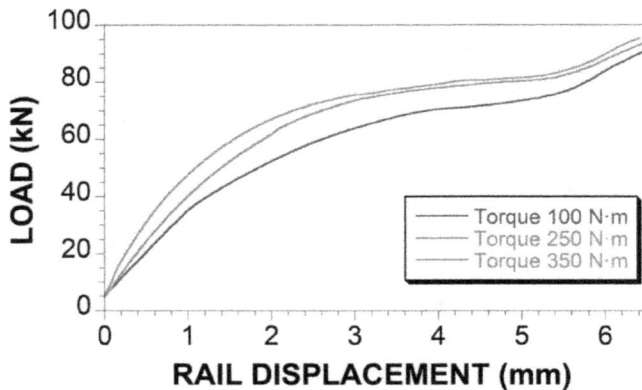

Figure 3.
Influence of the tightening torque on the static behavior of the plate.

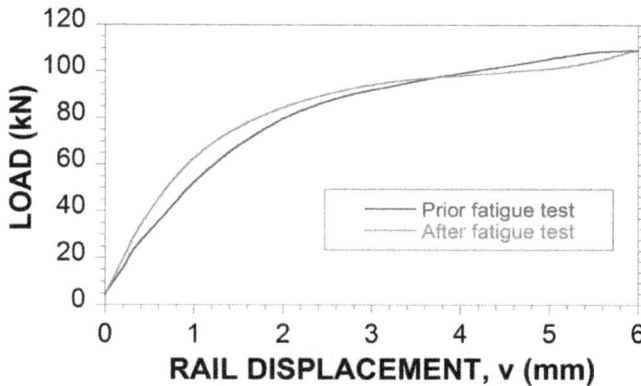

Figure 4.
Influence of fatigue aging on the static performance of the plates.

more easily with respect to the part that transmits the force. Sliding becomes more difficult as this contact force increases (when the tightening torque is higher).

2.1.3. Influence of fatigue deterioration on the static behavior

The assessment of the influence of fatigue deterioration on the static behavior was carried out on plates with the same humidity, 0.6%, one of them in the original state and the other deteriorated after being subjected to fatigue loading with the following parameters: frequency: 5 Hz, maximum force: 37.7 kN, minimum force: 2.5 kN, and the number of cycles: 500,000. The results can be seen in **Figure 4**, which represents the evolution of these two static tests prior and after fatigue. The behavior of the two plates is similar; the most remarkable difference consists in the greater initial stiffness of the previously fatigued plate; besides, no differences in the values of the contact load can be seen.

2.2 Impact resistance

As in the previous section, the goal of the tests is to determine the influence of variables such as moisture content, tightening torque or fatigue deterioration on the impact resistance of the component. The impact tests were performed applying square wave cycles under control of displacement conditions with sufficient amplitude to provoke the failure of the plate.

2.2.1 Influence of the moisture content on the impact resistance

To study the influence of the moisture content on the behavior under an impact loading, plates with the following humidity contents were used: dried in an oven, 0.5, 1.8, 2.2, and 3.6%. The results obtained after the impact tests of plates with different degrees of humidity are represented in **Figure 5**. The first remarkable difference with the static response is that the failure of the plates occurs in all cases before the contact with the screw. Only the plate with a moisture content of 3.6% may raise a doubt because its displacement slightly exceeds the limit of 4.4 mm established previously; nevertheless, since there is no change in the slope of the curve nor any type of mark on the plate after test, the contact with the screw can be discarded.

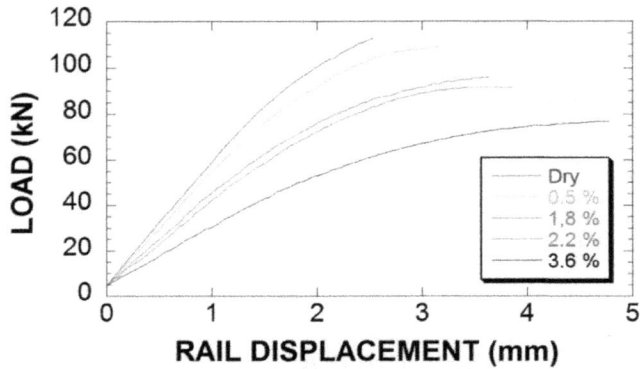

Figure 5.
Results of the impact tests on plates with different moisture contents.

The above figure makes it possible to draw the following additional conclusions: the stiffness of the plates increases as the moisture content decreases; notice that this pattern was also observed for the static tests. Moreover, the deformation at fracture increases with humidity, whereas the maximum force decreases.

2.2.2 Influence of the tightening torque on the impact resistance

To study the influence of the tightening torque, the same tests were carried out as in the static study, that is, the three following tightening torques were applied on dry plates: 100, 250, and 350 N·m, respectively. In **Figure 6**, the force-displacement curves of the three tests are shown. As can be seen, the plates show a stiffer behavior when the tightening torque is increased (as occurred for the static behavior); moreover, increasing the torque also increases the maximum force and decreases the deformation. This can be explained on the basis that, as the torque decreases, the contact force on the plate decreases too, facilitating the sliding of the plate and therefore its displacement.

2.2.3 Influence of fatigue on the impact resistance

To finish the study on the impact resistance, the influence of the deterioration due to fatigue was studied. The experiment was carried out on plates with a

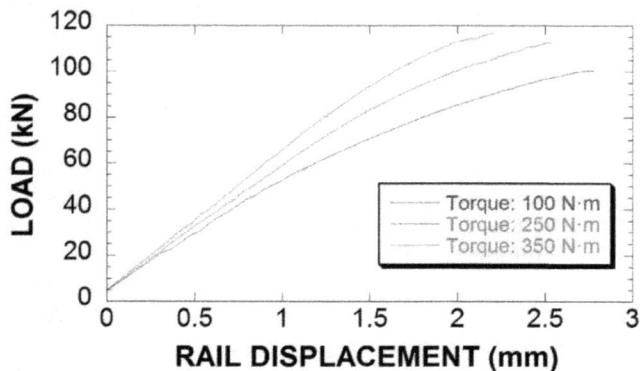

Figure 6.
Influence of the tightening torque on the impact behavior of the plate.

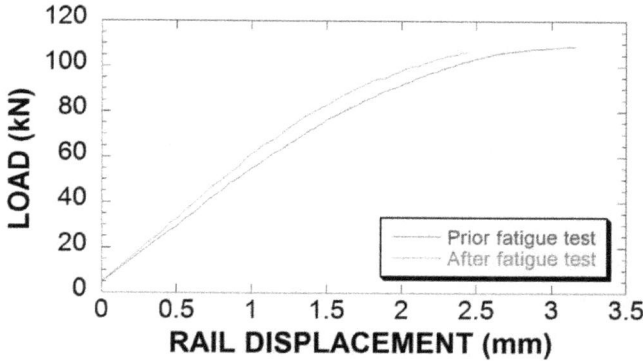

Figure 7.
Influence of the fatigue aging on the impact resistance.

humidity of 0.6% that were subjected to 500,000 sinus-type cycles with the parameters described above. The results are represented in **Figure 7**, where the load-displacement curves of two impact tests are shown, one performed on an original plate and the other on a fatigue-aged plate. The graph shows that fatigue aging by fatigue tends to stiffen the plate, decreasing the deformability at fracture by more than 0.5 mm without affecting the fracture load.

2.3 Fatigue behavior

In this section, the behavior of the plate subjected to dynamic cyclical loads will be studied. In addition, the influence on this behavior of external parameters, such as the moisture content of the plate or the value of the tightening torque applied to the clamping assembly will be assessed. For the fatigue characterization, the Locati test technique [3–5] was used; this consists in applying blocks (composed of a constant number of cycles) of increasing maximum load, starting from a value lower than the fatigue limit. At a certain level of maximum load, the maximum strain grows rapidly, preceding the failure of the plate.

2.3.1 Influence of the moisture content on the fatigue behavior of the plate

For this analysis, the Locati method was used applying loading blocks of 20,000 square wave cycles with a frequency of 5 Hz; the load for the initial block ranged between 5 and 45 kN. The minimum value was kept constant throughout the whole test, while the maximum load value was increased by 2 kN in each block. **Table 1** and **Figure 8** show, as a summary, the results obtained in the Locati tests carried out

Humidity (%)	Critical block	$\Delta\sigma_e$ (kN)	T_c (°C)	Cycles to failure
3.6	2	40	42	98,887
2.2	3	42	44	119,508
1.8	5	46	46	129,057
0.6	6/7	48	48.5	153,929
0	10/11	56	52	229,297

Table 1.
Influence of the moisture content on the fatigue behavior of the plates.

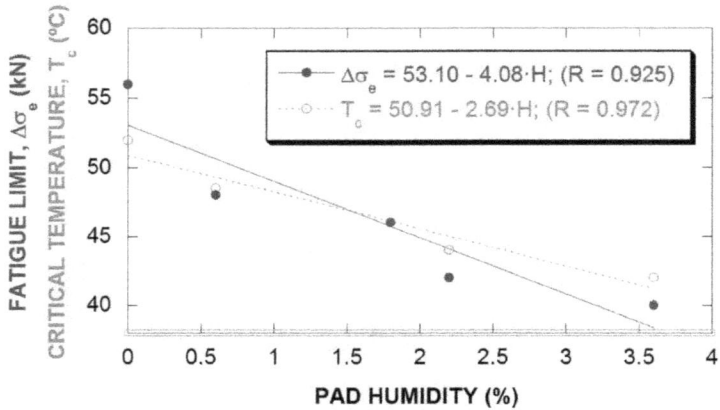

Figure 8.
Influence of moisture content on fatigue limit and critical temperature.

on plates with different moisture contents, verifying how the fatigue limit, $\Delta\sigma_e$, increases with the decrease in the moisture content of the plate. In the same figure, the evolution of the critical temperature, T_c, for the different moisture contents is represented, obtaining the same decreasing relation.

The dry plates under static loading failed in a brittle manner, as opposed to the ductile failure of the wet plates. In contrast, under impact conditions, both types of plates exhibit a brittle response. Finally, the fracture of the plates under fatigue loading is ductile in all cases, regardless of the level of moisture content.

2.3.2 Influence of the tightening torque on the fatigue behavior of the plate

As in the static and impact cases, the tightening torque applied in the fastening system may modify the fatigue behavior of the plate. To verify the influence of this parameter, Locati tests were carried out on plates with the same moisture content (0.6%) varying the tightening torque according to the following values: 100, 250, and 350 N·m. The parameters chosen in the Locati tests were the same as those described before. The increase in the tightening torque improves the fatigue behavior of the plate. **Table 2** and **Figure 9** show, as a summary, the results obtained in the Locati tests carried out on plates with different tightening torques and moisture contents.

As can be seen, as far as fatigue is concerned, increasing the tightening torque has similar consequences as reducing the moisture content. It can be concluded that the critical temperature depends not only on the material but also on the certain boundary conditions, in this case, on the tightening torque of the assembly.

Torque (N·m)	$\Delta\sigma_c$ (kN)	T_c (°C)	Critical block	Cycles to failure
100	44	41.5	60,000-80,000	107,832
250	48	48.5	100,000-140,000	153,929
350	52	52	140,000-180,000	196,377

Table 2.
Influence of tightening torque on the fatigue behavior.

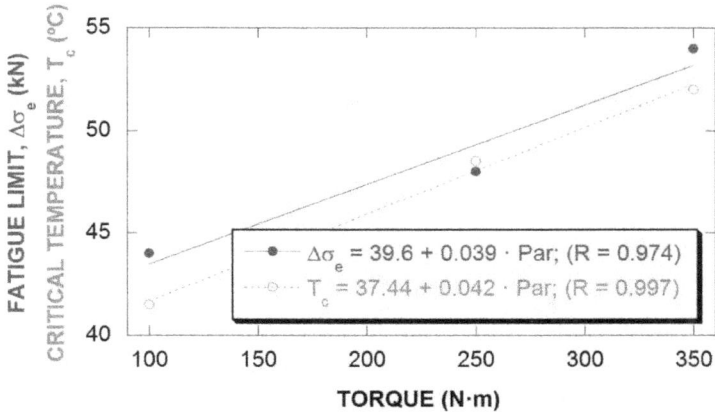

Figure 9.
Influence of the tightening torque on the fatigue behavior of the plate.

3. Rail pad

The rail pads used were made of thermoplastic elastomer (TPE) with a nominal stiffness of 100 kN/mm. As a new track is subjected to the repetitive loads caused by the passage of the trains, it suffers plastic deformations that grow until the system reaches the steady state and the material responds in a true elastic manner.

It has been verified [6, 7] that a high value of the stiffness of the rail pad increases the dynamic overloads due to the nonsuspended masses, accelerating the deterioration of the track, while a low value causes an excessive displacement of the track with a significant increase in the rail stresses. Therefore, it is necessary to delimit the validity interval of the stiffness. In the technical specification for the supply of fasteners of the GIF [8], both static and dynamic stiffness values are delimited. The value of the static vertical stiffness, k_s, must be included in the interval $80 \leq k_s \leq 125$ kN/mm, while the dynamic stiffness value, k_d, must belong to the interval $k_s \leq k_d \leq 2\,k_s$. Due to the nature of the constituent material of the pads, stiffness may be modified by different environmental agents such as temperature, which can fluctuate in track between –20 and 80°C, humidity or deterioration suffered by the pads due to the continued mechanical stresses of compressive fatigue [9]. In this section, the influence of these variables on the stiffness of the pads has been measured, in order to identify the conditions leading to the nonfulfillment of the above-mentioned requirements.

3.1 Influence of the temperature on the static behavior

The evaluation of the static behavior was determined from static stiffness tests (20/95 kN), applying a vertical load to the rail pad by means of the specific device. This device simulates the in-service working conditions, where the load was applied by means of a rail coupon, equipped with a ball joint that ensured the verticality of the loads. The mean vertical descent of the rail with respect to the support tool registered by four LVDTs (range: ± 5 mm), located in each of the corners of the support, was considered as the strain index of the rail pad. The load was applied by means of an actuator with a loading capacity of ±250 kN. The influence of temperature on the behavior of the rail pad was evaluated by introducing the device into an environmental chamber adapted to the test machine. Measurements of Shore D hardness [10] were carried out, aimed at finding a correlation of this parameter with the evolution of the mechanical behavior of the rail pad. The vertical static stiffness tests were carried out according to the provisions indicated in the technical specification [8] using a clamping force of 20 kN. The static behavior of rail pads at different temperatures (–10, 20, 50, and 80°C) was studied. In the graph of **Figure 10**, the evolution of the shortening of the rail pad in the third load-unload cycle at temperatures of –10, 20, 50, and 80°C can be seen.

The rail pad shortening ranges from maximum values of 0.93 mm at 80°C to a minimum of 0.59 mm at –10°C.

In **Figure 11**, the stiffness and the hardness of the rail pad versus temperature are represented in the double axis. An inverse linear correlation between static stiffness (20/95 kN) and the Shore D hardness with the rail pad temperature is shown; in summary, an increase in temperature causes the softening of the rail pad. An increase of 90°C in the test temperature, from –10°C, generates a 44% decrease in the stiffness of the rail pad (from 128 kN/mm to –10°C to 80 kN/mm at 80°C). The decrease in Shore D hardness is less noticeable (from 43.6 to 39.6) which is a decrease of 9.5% compared to the room temperature. The stiffness value is between

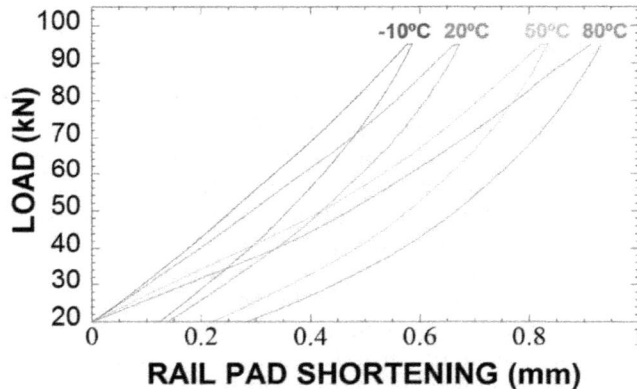

Figure 10.
Static behavior of the rail pad at different temperatures.

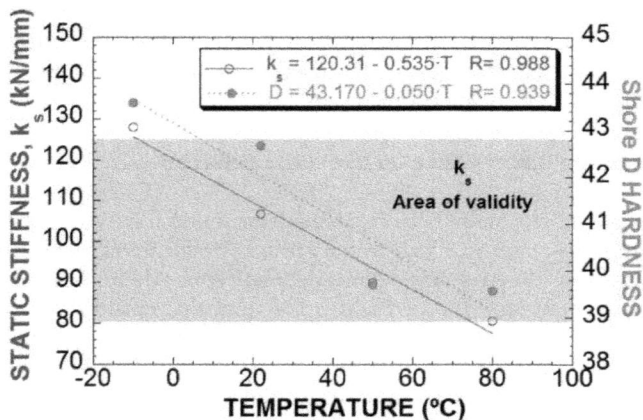

Figure 11.
Evolution of the stiffness and hardness of the rail pad with the temperature.

the established limits (80–125 kN/mm), except for the extreme temperature values
128.0 kN/mm to –10°C and 80.6 kN/mm to 80°C.

3.2 Influence of temperature on the dynamic behavior

The influence of temperature on the dynamic behavior was measured through
dynamic stiffness tests at low frequency. The test consists in applying 1000 load
cycles of a sinusoidal nature between 20 and 95 kN at a frequency of 5 Hz, deter-
mining the stiffness as the average obtained in the last 10 cycles. The dynamic
stiffness was analyzed at different temperatures: 20, 40, 60, and 80°C. In **Figure 12**,
the evolution of the shortening of the rail pad during the last cycle is represented.
The differences with respect to the previous static behavior are appreciated. Thus,
the rail pad only shortens by 0.64 mm at 80°C under dynamic conditions, having
reached a 31.2% higher static strain. This decrease in the strains also reflects a
decrease in the dynamic energy dissipated (E_{dd}); thus, the material under dynamic
conditions resembles an elastic body that does not dissipate energy (**Figure 13**).

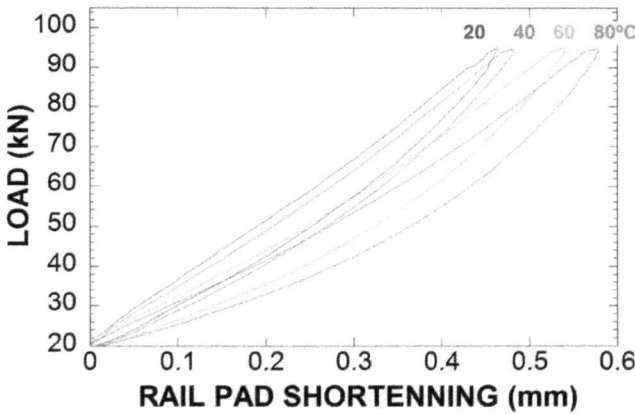

Figure 12.
Dynamic behavior at different temperatures (cycle 1000).

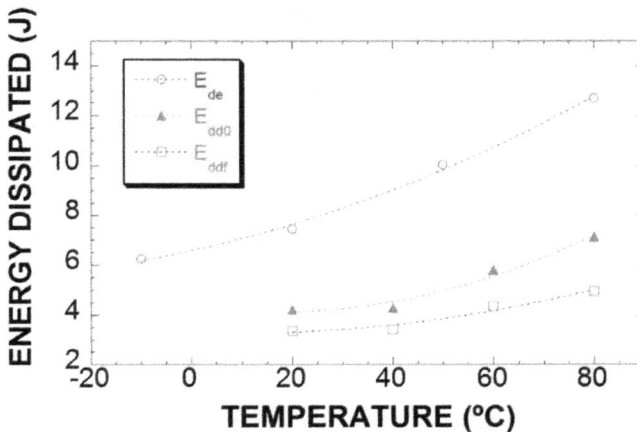

Figure 13.
Energy dissipated under static conditions (E_{de}) and dynamic (E_{dd}).

4. Anchorage components

The study of the anchor was separated into two parts: on the one hand, the polymer dowel was analyzed, studying its mechanical behavior under static and dynamic regimes; on the other hand, the metallic screw was subjected to loading configurations (tensile, bending, and impact) similar to those suffered under actual in-service conditions.

4.1 Dowel

The dowel consists of two components, namely, the main body made of polyamide 6.6 reinforced with glass fiber (30% wt.) and the metal sheet that surrounds the core peripherally. Due to the hygroscopic nature of the polyamide, the following treatments were applied to obtain a series of dowels with different moisture contents in order to know the influence of this parameter on the mechanical behavior:

- Drying in an oven at 100°C for 1 year.

- Maintaining the laboratory environment for 1 year.

- Immersion in water at room temperature for 3 months and then maintained in the laboratory environment for 9 months.

- Immersion in water at room temperature for a year.

These types of treatments were aimed at simulating the worst conditions that may take place in the preparations of the sleepers.

The moisture contents obtained with the different treatments were determined in two different ways. On the one hand, the overall content was measured weighing the dowel before and after the treatment and, on the other hand, the moisture content of the thread was calculated once it had been removed from the dowel in the mechanical tests and weighing the same before and after placing it in an oven for 7 days at 100°C. The latter procedure provides more representative values, since it is being measured in the resistant zone of the component. In addition, the humidity of the thread is, in all cases, higher than the overall humidity, since it is an area with greater external surface, and the water is absorbed more easily, being lighter in the last two treatments, since the interior area is the one that was in direct contact with the water that contained the dowel. The results obtained from the different moisture contents are shown in **Table 3**.

The critical stresses that the anchor must support are, mainly, parallel to the axis of the dowel-screw assembly, that is to say, forces that pull the screw out of the dowel. Therefore, the tool designed for the mechanical characterization of the dowels

Treatment	Total moisture (%)	Moisture thread (%)
1 year, dry in stove	0.00	0.00
1 year, environment	1.08	1.54
3 months, submerged in water	1.43	1.90
1 year, submerged in water	2.60	4.80

Table 3.
Humidity of the dowel according to the treatment.

LOAD (kN)

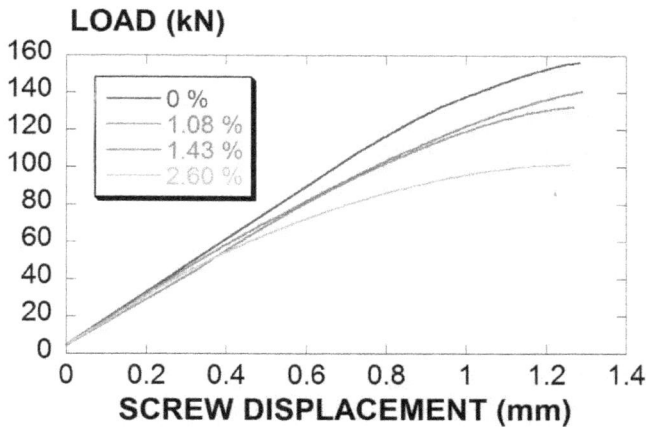

Figure 14.
Static test on dowels.

simulated this type of loading conditions. The characterization was based on both static and dynamic regimes, where the last one includes impact and fatigue efforts.

4.1.1 Characterization of the dowels under static loading

For the static characterization of the dowels, the load screw was driven to a depth of 64 mm, similar to the in-service position; then, an extraction force was applied until fracture at a loading rate of 1 kN/s. The results obtained on the dowels subjected to the different moisture treatments are represented in **Figure 14**. In view of the results, it is verified that the breaking load, as well as the initial stiffness of the component, increases as humidity decreases, while the deformation at fracture remains approximately constant. Therefore, the energy at failure decreases with increasing humidity. The constant value of the displacement at failure is explained because the applied stress is distributed uniformly along 14 thread passages, provoking the fracture of all of them simultaneously by shearing. As can also be seen, even for the highest moisture content, the limit value imposed by the technical specification (60 kN) is exceeded.

4.1.2 Characterization of the dowel under impact loading

In this case, the screw was driven to a depth of 64 mm and then a square wave cycle was applied to provoke the fracture of the component. The results obtained for the dowels subjected to the different moisture treatments appear in the graph of **Figure 15**, and these values are higher than those obtained in the static tests, while deformation increases with humidity, until reaching values similar to those obtained under static conditions. In the case of the two dowels with the lowest moisture content, fracture occurred at the depth where the screw thread reaches, this being the reason that explains why the fracture deformations of the dry samples were slightly lower than those of the wet ones.

The results allow it to be verified that the fracture load increases as the humidity decreases, characterization of the dowels under fatigue loading.

The Locati methodology was employed for fatigue characterization. Blocks of 20,000 square wave cycles were applied at a frequency of 5 Hz, with an initial loading range between 5 and 50 kN. The minimum load was constant during the

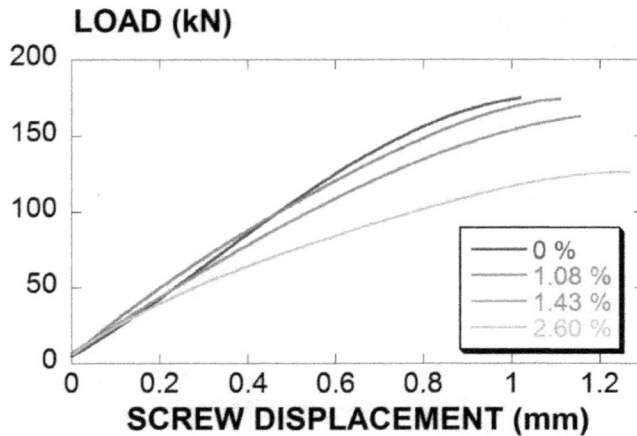

Figure 15.
Impact test on dowels.

test, while the maximum value was increased by 2 kN in each block. The results obtained in this experiment cannot be used to be compared with different tests because a nonquantified relative displacement between the dowel and the screw took place. The analysis of the displacement variation versus the number of cycles was used to determine the critical block. The results summarized in **Table 4** are obtained from dowels with different moisture contents in order to determine the critical parameters of fatigue.

The strong influence of moisture content on the fatigue behavior of this material is verified; note that the fatigue resistance of the dry dowel changes from 74 to 52 kN in the case of the wet dowel. The requirement of the technical specification regarding the extraction of fasteners is 60 kN in static regime; therefore, the fatigue limit exceeds this condition in all cases but for the case of maximum humidity.

4.2 Screw

The mechanical characterization was carried out following the requirements indicated by the internal control standard of the supplier [11]. The properties characterized were tensile, bending, and impact strength. Screws were used in the as-received state and were subjected to accelerated corrosion for 300 hours in a salt spray chamber to evaluate the influence of the exposure to a corrosive environment on the mechanical properties.

Humidity (%)	Critical Step	Critical F. (kN)	$\Delta\sigma_e$ (kN)	Critial number of cycles	Breakage nº of cycles
0/0	14	76	74	$260\text{-}280\cdot10^3$	321,101
1.08/1.54	12	72	70	$220\text{-}240\cdot10^3$	281,961
1.43/1.90	10	68	66	$180\text{-}200\cdot10^3$	251,072
2.60/4.80	3	54	52	$40\text{-}60\cdot10^3$	87,989

Table 4.
Summary of the results obtained in the Locati tests.

4.2.1 Tensile characterization

For the tensile characterization, a specific setup was designed and manufactured. Deformations were recorded by means of an extensometer with a gauge of 12.5 mm placed on two of the fillets of the screw. This layout allows the possible failure of the head of the screw to be assessed. Tests were carried out on a screw in the as-received condition as well as on two of the screws previously submitted to the salt spray chamber, one of which had even suffered a small notch in the bottom of a fillet, in order to accelerate the corrosion process. The graph of **Figure 16** shows the results of the tensile tests performed on the three samples, and the parameters obtained in each of the tests are summarized in **Table 5**. As can be seen, the effect of exposure in the salt spray chamber has not penalized the tensile mechanical behavior of the screw. Even the notch has had little influence, since the failure occurred in a different cross section.

4.2.2 Bending characterization

The bending test consisted in obtaining, respectively, a permanent angle of 15 and 30°, verifying the integrity of the sample when bending over a 40 mm radius. As in the previous case, the test was carried out on a sample in the as-received condition and on another one that had remained 300 hours in a salt spray chamber. After applying a bending of 15°, no visible defects appeared in any of the screws. With the 30°, small cracks can be seen in both samples; consequently, no differences are appreciated between samples regarding the bending ability. **Figure 17** shows the appearance of the samples tested after bending 30°.

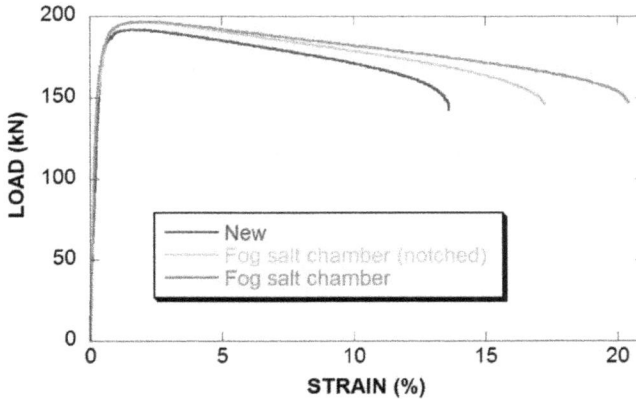

Figure 16.
Tensile test on the screws.

Screw	Maximum force (kN)	Strain under Max. Force (%)	Cross section reduction (%)
New	191.7	1.80	13.65
Fog salt chamber (notched)	196.2	1.96	14.13
Fog salt chamber	196.6	2.02	12.56

Table 5.
Parameters obtained in tensile tests.

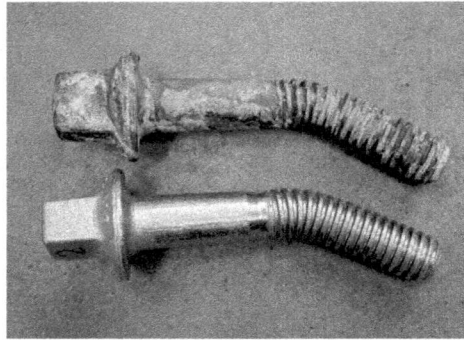

Figure 17.
Appearance of the samples after the bending test.

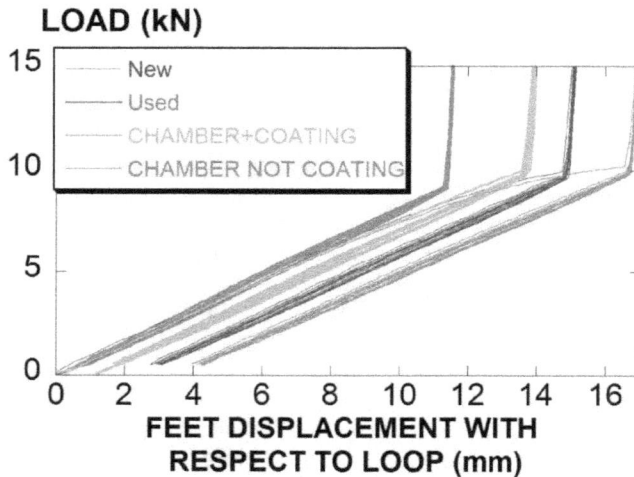

Figure 18.
Load cycles on clip.

5. SKL-1 clip

This part of the study was aimed at assessing the loss of mechanical performance of the clip SKL-1 depending on the degree of use. New clips, clips used under normal and extraordinary conditions, and clips subjected to different deterioration treatments were used. To simulate in-service conditions, a compressive load of 15 kN was applied for 10 s, and the sample was unloaded to 0.5 kN; this sequence was repeated nine more times. **Figure 18** shows the experimental curves obtained from the clips previously described. The comparison of responses was carried out using clips subjected to the following different conditions:

- New clip, not previously tightened.

- Used clip, tightened 22 times with a torque of 250 N·m and then three times with 350 N·m.

LOAD (kN)

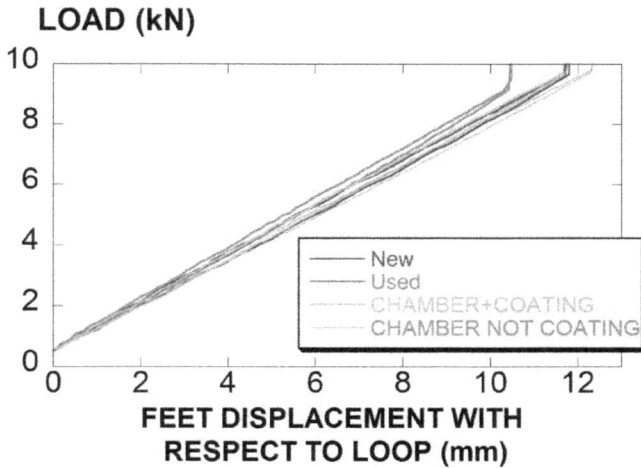

Figure 19.
Detail of the last load cycle on clip.

- Clip with coating introduced in a salt fog spray chamber for 300 hours and with a pretightening of 250 N·m.

- Uncoated clip inserted in salt fog spray chamber for 300 hours without tightening.

The difference between the first cycle and the rest of them was analyzed. The used clip was the only one in which this difference did not occur, since it behaved elastically from the first cycle. The clips that had not been previously tightened show a first cycle with higher compliance and permanent deformation. In addition, in the case of the salt chamber clip without coating, the flexibility is greater than in the new one, possibly due to the loss of net section because of the corrosion process. In all cases, the following cycles do not show plastic deformations. **Figure 19** shows a detail of the last 10 load cycles applied to the four clips.

The highest stiffness corresponds to the clip used, and the greatest flexibility to the clip subjected to the salt chamber without coating. As for the new clip and the salt chamber with coating, the difference between them is negligible, showing a behavior in an intermediate situation between the previous ones.

6. Conclusions

The increase in speed and the improvements in comfort and safety experienced by the high-speed train in recent decades are the result of the engineering innovations implemented in this means of transportation. In this chapter, the studies developed by the LADICIM research group on the design of the fastening system between the sleeper and the rail have been examined. Once verified that the fastening system selected for the high-speed line between Madrid and Barcelona fulfills the requirements of the European standards EN 13146, this contribution focuses on the experimental results derived from the characterization of each of its components. The conclusions drawn from this research are summarized hereafter:

- Guide plate

Increasing the moisture content reduces the strength and increases the deformability of the plate. This reduces the static force to be applied to provoke the contact with the screw. In case of an impact, the plate fails before contacting the screw, and fracture occurs with greater deformation and less force when the moisture content increases. This increase also causes a decrease in fatigue resistance.

The increase of the tightening torque applied to the system raises the force required to achieve the plate-screw contact under static conditions, reduces the impact deformability, and increases the resistance against fatigue.

A plate tested for 500,000 cycles has approximately the same behavior under static and impact conditions as an original plate.

- Seat pad

Temperature reduces the stiffness of the plate, both under quasi-static or dynamic conditions. The energy dissipated per cycle is accentuated by increasing the temperature.

Hardness correlates linearly with stiffness; then, hardness can be considered as an index of the degree of degradation undergone by the pad. This parameter is very useful as it can be easily measured by a nondestructive test on the track under in-service conditions.

Extreme temperature values on the railway track (–20 and 80°C) define the threshold working values for seat pads, both under dynamic and static conditions.

- Anchorage components

Increasing the humidity content of the dowel reduces the resistance under static, impact, and fatigue loading. Even in the worst scenario, this component fulfills the minimum requirements imposed by the European standard.

The screw subjected to saline fog chamber for 300 h shows a tensile behavior very similar to an original screw; the presence of a small notch does not reduce its mechanical resistance. In both cases, the screw survived to the bending test.

- Clip

The only difference between a used clip, a new one or one subjected to 300 h in a saline fog chamber lies in the first loading cycle; once the component is plasticized, the behavior is similar.

Author details

Estela Ruiz, Isidro A. Carrascal, Diego Ferreño*, José A. Casado and Soraya Diego
LADICIM (Laboratory of Science and Engineering of Materials), University of
Cantabria, E.T.S. de Ingenieros de Caminos, Canales y Puertos, Santander, Spain

*Address all correspondence to: diego.ferreno@unican.es

IntechOpen

References

[1] Carrascal IA. Optimización y análisis de comportamiento de sistemas de sujeción para vías de ferrocarril de alta velocidad española [thesis]. Santander: University of Cantabria; 2010

[2] Soluciones M. Fastening systems [Report]. 2009

[3] Locati L. La Fatica dei Materiali Metallici. Milano: Ulrico Hoepli; 1950

[4] Locati L. Programmed fatigue test, variable amplitude rotat. La Metallurgia Italiana. 1952;**44**(4):135-144

[5] Casado JA, Polanco JA, Carrascal I, Gutiérrez-Solana F. Application of the locati method to material selection for reinforced polymeric parts subjected to fatigue. In: International Conference on Fatigue of Composites. Paris. Proceedings of the Eighth International Spring Meeting. June 1997. Issue 8. pp. 454-461. http://www.worldcat.org/title/international-conference-on-fatigue-of-composites-conference-internationale-sur-la-fatigue-des-composites-eight-international-spring-meeting-huitiemes-journees-internationales-de-printemps-paris-3-4-5-june-1997/oclc/496942402

[6] Lopez Pita A. La rigidez vertical de la vía y el deterioro de las líneas de alta velocidad. Revista de obras públicas; November 2001. Issue 3415. pp. 7-26. http://ropdigital.ciccp.es/revista_op/detalle_articulo.php?registro=18223&anio=2001&numero_revista=3415

[7] Carrascal IA, Casado JA, Polanco JA, Gutiérrez-Solana F. Dynamic behaviour of railway fastening setting pads. Engineering Failure Analysis. 2007; **14**(2):364-373

[8] EN13481-2:2012+A1:2017, Railway applications. Track. Performance requirements for fastening systems.

Part 2: Fastening systems for concrete sleepers, and EN 13481-5:2012+A1:2017, Railway applications. Track. Performance requirements for fastening systems. Part 5: Fastening systems for slab track with rail on the surface or rail embedded in a channel

[9] Carrascal IA, Casado JA, Polanco JA, Gutiérrez-Solana F. Comportamiento dinámico de placas de asiento de sujeción de vía de ferrocarril. Anales de mecánica de la fractura. Vol. XXII. 2005. Almagro. http://www.gef.es/web/Publicaciones.asp

[10] UNE-EN ISO 868. Plásticos y ebonitas. Determinación de la dureza de indentación por medio de un durómetro. (Dureza Shore); 2003. https://books.google.es/books/about/UNE_EN_ISO_868_2003_pl%C3%A1sticos_y_ebonita.html?id=F46MswEACAAJ&redir_esc=y

[11] Technical Specification ET 03.360.572.6 B080805, Anclaje tipo Plastirail. ADIF (Spanish Administrator of Railway Infrastructures)

Chapter 3

Contemporary Inspection and Monitoring for High-Speed Rail System

Lu Zhou, Xiao-Zhou Liu and Yi-Qing Ni

Abstract

Non-destructive testing (NDT) techniques have been explored and extensively utilised to help maintaining safety operation and improving ride comfort of the rail system. As an ascension of NDT techniques, the structural health monitoring (SHM) brings a new era of real-time condition assessment of rail system without interrupting train service, which is significantly meaningful to high-speed rail (HSR). This chapter first gives a review of NDT techniques of wheels and rails, followed by the recent applications of SHM on HSR enabled by a combination of advanced sensing technologies using optical fibre, piezoelectric and other smart sensors for on-board and online monitoring of the railway system from vehicles to rail infrastructure. An introduction of research frontier and development direction of SHM on HSR is provided subsequently concerning both sensing accuracy and efficiency, through cutting-edge data-driven analytic studies embracing such as wireless sensing and compressive sensing, which answer for the big data's call brought by the new age of this transport.

Keywords: non-destructive testing, structural health monitoring, defect detection, fibre Bragg grating, high-speed rail, sensing technology

1. Introduction

The past decade has witnessed the most prosperous blooming of HSR, marking a splendid new age of this fast-developing transportation, which subtly alters people's travelling habit with great convenience and ride comfort. Hidden behind the high-quality ride service provided by HSR is the tremendous effort and huge budget spent on the inspection and maintenance work, which is more challenging with increasing speed and capacity.

With long-term numerous cycles of loading and unloading, both rail tracks and train wheels are suffering from vibrations and stresses caused by wheel/rail inter-actions, leading to fatigue, wear, plastic deformation, cracks and other deteriorations. The wheel/rail interactions are intense with average contact stresses over 1000 MPa under normal operating conditions, and this number can go much higher upon specific situations (wheel flange/rail edge contact while train turning, poor conforming wheel and rail profiles, etc.) [1]. Moreover, to author's knowledge with recent research work on contact mechanics using NDT approaches, machine element contacts including wheel/rail contacts are essentially contacts between the asperities due to surface roughness of the contact bodies, and the asperity contacts indicate hyper-stress concentration beyond 4000 MPa at the contacting peaks [2]. Under such high stresses, components of the rail system are deteriorating rapidly in various forms and the deteriorated structures create a worse operating environment, adding the occurrences of failures. A typical failure is rolling contact fatigue (RCF) causing a series of subsequent rail defects (squats, transverse cracks, spalling and gauge corner cracks).

The rail also takes up impact load from running trains intermittently due to wheel defects, rail irregularities or at certain areas rail turnouts, rail joints, etc. The intense vibrations caused by wheel/rail interactions and impacts are transmitted bidirectionally from the wheel/rail interface up to the coach and down to the rail slab simultaneously. In terms of HSR, to meet the high standard requirements of smooth operation under high speed, the components utilised are different from those in conventional rail lines. For example, the rail tracks are strengthened with high resistance to wear, and multi-layer concrete forms up the rail slab with CA mortar layer serving as the damping instead of traditional ballast. These measures add ride comfort in HSR operation, but make the system more 'brittle' with reduced capability in vibration absorption, hence add the risks of cracks in the rail system. A recent example is the giant crack (44 cm long) found in an operating Japan Shinkansen bullet train in December 2017, causing interruption of service and great social panic [3]. Similar cases can be highly possible on ballastless rail tracks leading to more catastrophic consequences, calling for more reliable and thorough inspection actions.

NDT techniques have long been used for inspection in rail system since the 1920s. With integrated ultrasonic probes or eddy current sensors, the NDT systems are able to check surface and internal defects along the rail in either contact or non-contact manner. The NDT inspection is conducted through manual inspection device or inspection vehicle. Conventional inspection vehicles are normally attached to a traction locomotive to carry out inspection. In the age of HSR, many countries have developed high-speed comprehensive inspection vehicles (CIVs) for the more complicated inspection tasks, such as the 'East-i' CIV in Japan, the 'IRIS320' CIV in France, and the 'No. 0' CIV in China, etc. Inspection content of the high-speed CIVs covers from geometry data of rail infrastructure to dynamic behaviours of trains. Despite of the wide range of data types, the NDT techniques require interruption of train service to conduct the inspection. To provide early alarming in prevention of further consequences in terms of accidents similar to the Japan Shinkansen case, continuous real-time information of in-service rail system is highly desired, which puts forward the introduction of online monitoring to this area. Since wheel/rail interaction is the core part of the rail system, this chapter mainly focuses on the inspection and monitoring methods of wheel and rail defects.

2. Typical defects of wheels and rails

2.1 Wheel out-of-roundness (OOR)

Various types of wheel OOR/defects occur on HSR in-service, which influence operational safety and give rise to high maintenance cost. These defects take on many patterns, such as flats, eccentricities, polygons, corrugations on block-braked wheel treads, missing pieces of tread material owing contact to fatigue cracking and other random irregularities [1, 4]. Generally, they can be categorised into two major types: local defects and periodic OOR all around the wheel. The former can cause severe repeated wheel-rail impacts, while the latter leads to abnormal vibrations of vehicle-track system at certain frequencies [5].

2.1.1 Wheel local defects

There are two major causes behind initiation and development of wheel tread local defects: thermal cracking and rolling contact fatigue (RCF) [6]. Several factors, such as speed, axle load, wheel-rail adhesion, wheel material and braking conditions, also have some effects on deterioration rates of wheel tread [7]. In HSR operation, wheel wear rate can increase quickly due to the high operation speed, high stiffness track, wide wheel-rail impact frequency, intense vibrations and high speed flow [5, 7, 8]. Wheel defects can cause abnormal vibrations and have the potential to impose damage to both track and vehicle components such as sleepers, rails, wheelsets and bearings, increase the likelihood of derailment and deteriorate operational safety and comfort owing to high vibration amplitudes [1, 9]. Previous research found that the load history of axle bearing and bogie frame may fluctuate due to the influence of wheel roughness and lead to fatigue cracks [10]. Wheel defects also result in an increase in the noise both inside and outside the train [11, 12] which can be annoying for both passengers on the train and residents along the rail line [5]. For high-speed trains, the high-magnitude impact loads generated by a defective wheel can excite various vibration modes for the wheelsets and thereby contribute to abnormal increases in the stress states of wheel axle under high-speed conditions [13].

2.1.2 Wheel polygonisation

The studies of wheel polygonisation were stated some three decades ago when some of polygonal wheels were detected on high-speed trains (ICE, Germany). Wheel polygonisation with one, three and four harmonics around circumference has been found on disc-braked wheels in ICE, in which the third harmonic dominated for solid steel wheels, while the second harmonic was common for rubber sprung wheels [5]. The research on high-order polygonisation (15–25 orders) had not been carried out until recent years, when new problems and challenges in HSR operation were raised. For HSR, there is an increasing demand for relative studies on this problem because it is reported that high-order polygonisation with very small radial deviation (< 0.05 mm, or <20 dB re 1 µm) can cause abnormal vibration and even failures to the bogie components. The influences of polygonal wheels on track structure and vehicle components are studied by [13, 14]. It is revealed that: (1) the wheel-rail impact normal force increases with the deepening of the wheel polygonal wear; (2) the amplitude of the normal force fluctuation depends mainly on the wavelength and depth of the wheel polygonal wear on the wheel running surface; and (3) the stress load cycles induced by wheel polygonisation can considerably increase the propagations of the initial crack in the wheel axle.

2.2 Rail defects

2.2.1 Common defects caused by inappropriate manufacturing and use

As per increasing of demand for HSR, rail defects have become a critical challenge in operation because an incident could cause more losses when trains run at higher speed. Many researchers have proposed classification methods for typical types of rail cracks derived from different propagation orientations of rail defects [15, 16]. The most common rail defects are caused by inappropriate manufacturing and inappropriate use of rails, and they mainly include transverse defects (TD), detail fractures (DF) and split heads. TD (**Figure 1a**) is one of the most critical type

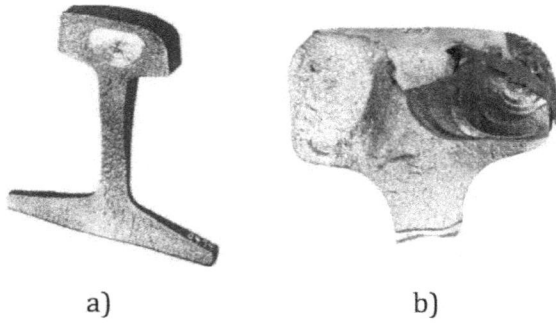

a)   b)

Figure 1.
Transverse defect and detail fracture [15]. (a) Transverse defect. (b) Detail fracture.

of cracks that appear in railheads propagating along lateral direction. DF (**Figure 1b**) has an origination point and grows radially from the origination point. These types of defects are caused by inappropriate use such as excessive stress concentrations. The vertical split heads (VSH) (**Figure 2a**), which usually originate from manufacturing anomalies, cause the second most train derailments (after TD).

2.2.2 Rolling contact failure (RCF)

RCF has become a significant economic and safety challenge for HSR and metro lines. Fatigue fracture occurs as a result of a periodic loading applied to the materials which exceeds its fatigue limit. Normally, it will lie between 35 and 60% of the tensile strength of rail [17]. Once a fatigue crack has initiated, it will spawn with every period of loading. At the beginning, it is very gentle and then quicker until a critical size is achieved [18]. Typical RCF originating at rail surface includes head checks, surface gauge corner head checks and squats. The cracks generate as the rails experience huge impacts from the wheels [19, 20] and the fatigue damage results from the normal and shearing stresses of the wheel-rail interaction [21]. A micro-crack may induce surface spalling effect when it propagates from the rail-head to inner parts of the rail. In addition, RCF can cause corrugation and bolt hole cracks on the rails, significantly influencing track structure. Generally, there are six classes of corrugation: short-pitch corrugation, light rail corrugation, corrugation on sleepers, contact fatigue corrugation, rutting and roaring rails and heavy haul corrugation [22].

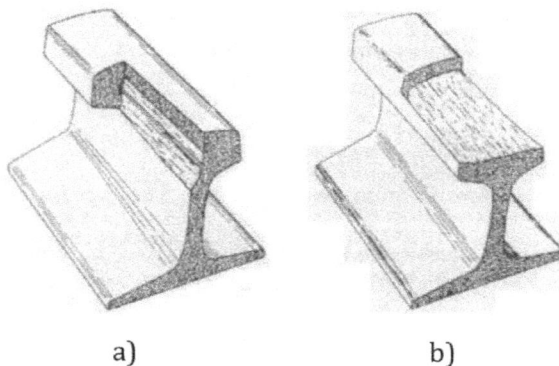

a)   b)

Figure 2.
Split head [16]. (a) Vertical split head. (b) Horizontal split head.

3. Wheel roughness measurement and defect detection

The most effective and common strategy to control the wheel defects is wheel re-profiling [5] which can eliminate local defects and polygonisation and reduce the resulting noise and vibration [4]. In modern HSR wheel maintenance, many modern depots are equipped with a wheel re-profiling facility known as a wheel lathe and the wheelsets are not necessary to be disassembled during re-profiling. However, the wheel re-profiling always follows a time or mileage-base schedule per earlier experience or supplier's specification. Consequently, it can decrease the wheel diameter and thereby shorten the service lives of the healthy wheels which are scheduled to be re-profiled. Therefore, there is a large economic incentive for adopting condition-based maintenance (CBM) scheme based on advanced NDT and SHM techniques, to reduce maintenance costs of wheelsets and efficiently preventing the hazards imposed by wheel defects. There are two main types of CBM approaches: in-service (online) condition monitoring and in-depot (offline) inspection [23]. The former one provides real-time condition information for maintenance planning, while the latter approach, normally done at a fixed interval, can offer accurate measurement for condition assessment of vehicle components.

3.1 Wheel roughness measurement/in-depot (offline) wheel inspection

Wheel tread roughness measurement (in-depot inspection) is a direct way of collecting wheel condition information for maintenance, monitoring profiles in conjunction with wear problems. With wheel roughness measurement data, the wheel re-profiling strategy can be optimised using data-driven wear model [24]. The NDT technologies employed by roughness measurement include linear variable differential transformer (LVDT), the mechanical displacement probe, the rotation sensor, electromagnetic acoustic transducer (EMAT) [25], laser-ultrasonics [26], laser-air hybrid ultrasonic technique (LAHUT) [27] and other novel NDT techniques [28–30]. Some even allow the trains run at a low speed during inspection [30]. However, for measurement methods using ultrasound pulse-echo technique, it is sometimes difficult to detect wheel flats because they usually have smooth edges that do not generate echoes [31]. There are now many commercial devices that allow the measurement to be done in depot with high efficiency, such as ØDS measurement instrument, Miniprof, MÜLLER-BBM, etc.

3.2 Online wheel condition monitoring and defect detection

Existing online wheel condition monitoring systems mainly include trackside wheel impact load detector (WILD), force gauges installed on sleeper pads, distributed sensors based on Brillouin optical time domain analysis (BOTDA), accelerometer-based trackside detector, acoustic detectors, laser- and video camera-based detectors, etc.

3.2.1 Trackside WILD and wheel: rail interaction detector

By deploying strain gauges and accelerometers on the rail, it is possible to measure wheel-rail contact force or rail acceleration response when a train passes over the instrumented rail section. These devices report impact as either a force at the wheel-rail contact interface or a relative measure of the defect [10]. The most common WILD is composed of a series of strain gauge load circuits mounted on the neutral axis of the rail between two adjacent fasteners in several consecutive sleeper bays to quantify the wheel-rail interaction force.

Johansson and Nielsen [1] made use of this set-up to build a detector on the rail web in nine consecutive sleeper bays. Nielsen and Oscarsson [32] used both rail web and rail foot strain gauges to measure the wheel impact load and rail bending moment. Stratman et al. [33] proposes new criteria for removal of wheels with high likelihood of failure, based on two real-time SHM trends that were developed using data collected from in-service trains. Filograno et al. [34] developed an FBG-based sensing system comprising FBG strain gauges mounted at both rail web and rail foot enables train identification, axle counting, speed and acceleration detection, wheel imperfections monitoring and dynamic load calculation. They have expanded the application of this system in the Madrid-Barcelona HSR line [34]. An FBG-based wheel imperfection detection system that can provide in-service measurement of wheel condition was developed by The Hong Kong Polytechnic. It offers a comprehensive health monitoring scheme for vehicle and track in the entire railway network of Hong Kong [35]. A monitoring system has been proposed with FBG sensors implemented on rail tracks to detect wheel local defects such as wheel flats and polygonisation [36]. The impacts of wheel/rail interactions caused by wheel local defects are reflected as subtle anomalies in response to signals collected by FBG sensors, and the deployed system is shown in **Figure 3a**. The detecting results match well with those from offline inspection (**Figure 3b**).

In addition to the strain gauge-based detector, there are other methods for online wheel load measurement to assess the condition of passing wheels, such as force gauges installed on sleeper pads, distributed sensors based on Brillouin optical time domain analysis (BOTDA), etc. Besides, there are also some commercial WILDs, such as WheelChex® system, GOTCHA system (optic fibre-based wheel-flat detection and axle load measurement system), and MULTIRAIL WheelScan.

3.2.2 Trackside rail acceleration and noise detector

Accelerometer-based systems can provide 100% coverage of the circumference of a wheel of any size in defect detection [10]. Skarlatos et al. [37] used two B&K accelerometers placed on the rail foot to pick up the rail vibration signals for diagnosis of wheel defects. Belotti et al. [38] used four consecutive accelerometers and an inductive axle-counter block which help to discriminate the response corresponding to each wheel. Seco et al. [39] proposed a trackside detector which has eight accelerometers installed on both bend zone and straight zone. However, the acceleration data are difficult to convert to wheel-rail impact load, which is widely used as wheel local defect indicator [5]. This is mainly because the measured acceleration signal could not directly refer to the excitation of each wheels, and

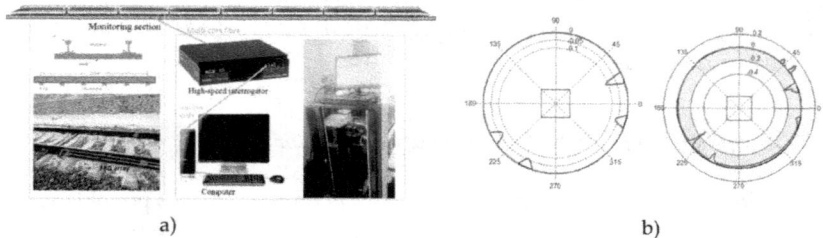

a)   b)

Figure 3.
High-speed train wheel defect detection using online FBG-based system [36]. (a) System configuration. (b) Detection results (left—online detection; right—offline inspection).

sometimes an additional axle counter is needed [23]. Furthermore, their performance might be limited by their repeatability and by the analysis applied to the accelerations acquired.

The commonly used noise detector is called trackside acoustic array detectors (TAADs), which make use of arrays of high-fidelity microphones to listen to the audible noises produced by the passing trains [40]. There are also some commercially available systems, such as trackside acoustic detection system (TADSTM) and the RailBAMTM [41, 42]. However, these systems are specialised in wheel bearing fault diagnosis rather than wheel tread defect detection. For the slight flat defect of high-speed train wheel (flat depth < 0.5 mm), this method may not be applicable because when the train runs at high speed (>200 km/h), the prediction accuracy can be limited [11].

3.2.3 Detection methods based on laser and video cameras

There are various types of wheel roughness monitoring systems based on laser and video cameras. Some typical or well-known detectors include wheel profile detectors (WPDs), MBV-systems, wheel profile measurement system (WPMS) and those based on light illumination devices, light-sensing devices, charge-coupled device (CCD) camera, and laser displacement sensors (LDSs).

WPDs are based on a combination of lasers and video cameras to automatically measure the wheel profile while train is in motion [40]. These data acquired from WPDs include wheel profile and wear, wheel diameter, height and thickness of the flange, back-to-back distance and wheel inclination. A prototype of a condition monitoring system called MBV-systems is presented by Lagnebäck [43] for measuring the profile of the wheels with a laser and a camera. An automatic WPMS based on laser and high-speed camera was installed on an iron ore line in Sweden in 2011 and can measure the wheel profile for speeds up to 140 km/h [44]. This system, which is in a CBM manner, has been attracting more and more interest from maintenance engineers in the Swedish railway sector. Zhang et al. [45] presented an online non-contact method for measuring a wheelset's geometric parameters based on the opto-electronic measuring technique. The system contains a charge-coupled device (CCD) camera with a selected optical lens and a frame grabber, which was used to capture the image of the light profile of the wheelset illuminated by a linear laser. Besides, there are some newly designed laser-based online detectors, which are immune to vibration and on-site noise, easy to calibrate, with high efficiency of data acquisition and with high accuracy of positioning [46, 47].

4. Rail defect detection and monitoring techniques

Manual inspection method is still widely used in most routine track inspections until today since it can directly figure out rail defects. However, it needs experienced workers and involves significant human input and judgement [48]. Therefore, NDT&E techniques, which enable rail inspection in an automated manner, are in need. NDT techniques were advocated for rail inspection as early as the 1920s [49]. Ultrasonic testing (UT) emerged in the 1960s became dominant in rail inspection [10, 50]. With the development of UT, European countries and Japan have released a variety of forms of ultrasonic rail flaw detection equipment, such as portable types, hand-push types, road and rail dual-use vehicles and specialised rail-testing trains [28]. While being extensively utilised, both the magnetic induction testing and conventional UT methods are not suitable for all defect scenarios; for example, they offer poor sensitivity to defects located in the rail web and rail foot [51]. A wide variety of inspection techniques are under research and development with the target to enhance the detection capability.

4.1 Advanced UT techniques

To enhance accuracy, speed and detection rate in rail defect detection, many research efforts have been made to improve the detection methods and develop advanced UT techniques. The novel techniques include laser ultrasonic testing (LUT), phased array ultrasonic testing (PAUT), electromagnetic acoustic testing or electromagnetic acoustic transducer (EMAT), guided wave testing (GWT) and acoustic emission testing (AET).

4.1.1 Laser ultrasonic testing (LUT)

Compared with traditional piezoelectric UT, LUT has its own merits such as non-contact and no coupling agent. The laser device can be located relatively far away from the rail with optic-fibre used as transmission media. This enables the establishment of trackside monitoring system. Besides, with good interference immunisation, laser can be used in measurement in adverse environment or high temperature. Pulsating laser works on solid surface and produces longitudinal wave, lateral wave and surface wave simultaneously. As a result, it can be applied to detect not only surface defects but also internal defects. Yet certain problems exist in LUT, such as low efficiency of light-sound energy transformation, weak ultrasonic signal and high cost of detection equipment.

Nielsen et al. [52] developed an automatic LUT-based system for rail inspection, named LURI, which was tested on a railroad line containing man-made structural defects. This system can detect defects on the running surface of the rail, as well as horizontal and vertical flaws in the railhead. Kenderian et al. [53] developed the first non-contact testing system based on laser-air hybrid ultrasonic technique for rail defect inspection. The system can detect VSH defects and thermal fatigue cracks with a success rate of nearly 100%, and rail web defects with a rate of approximately 90%. Lanza et al. [54] developed a laser/air-coupled rail defect detection system, which can accurately locate rail transverse cracks by using laser emission and ultrasonic wave for detection.

4.1.2 Phased array ultrasonic testing (PAUT)

PAUT, developed from the research on phased array radar, can detect cracks in different directions, depth and locations conveniently. Utrata and Clark [55] present groundwork of PAUT methods, which provided useful information and evidences for the positioning of phased array (PA) probes in rail flaw detection. PAUT is now widely applied in rail defect detection, covering railhead, rail web, rail base and weld areas. Institutes that carry out research on PAUT for rail defect detection include: Transportation Technology Centre Inc. (TCCI), Lowa State University, University of Warwick, University of Birmingham in UK, TWI Company and Socomate in France, etc.

Wooh and Wang [56] developed a hybrid array transducer which is an assembly of a linear phase and a static array and can accurately assess real defects in rail specimens. Speno International Company [57] developed an ultrasonic rail testing equipment based on multi-element phased array technology and the equipment was installed on a trial inspection car which can achieve a speed of 80 km/h and a sampling rate of 6 kHz when detecting rail defects. TTCI [58] developed an

Omni-scan PAUT system which was applied in on-site detection of TDs. Field test of the system was conducted on the Facility for Accelerated Service Testing (FAST) of TTCI.

4.1.3 Electromagnetic acoustic testing (EMAT)

EMAT, as a kind of excitation and detection technique of propagating ultrasonic wave, can provide detection of defects located in subsurface area of railhead. Thus, a promising method appears to be electromagnetic-acoustic method which is realised by EMAT transducers. Both transverse and longitudinal cracks in railhead can be detected by using EMATs, as shown in **Figure 4** [59]. University of Warwick and University of Birmingham [60] developed a railway surface-detect inspection technique based on EMAT equipped with two EMAT converters, one for emitting surface Rayleigh waves and the other for receiving surface propagating Rayleigh waves. It is found that this technique can improve the inspection rate of horizontal and vertical defects on the railheads, compared with piezoelectric transducers. University of Warwick [61] designed a lab-based laser-EMAT system to investigate the ultrasonic surface wave's generation, propagation and interaction on the rail-head with a Michelson interferometer measuring the out-of-plane displacement. The Rayleigh-like wave generated by EMAT can flood the whole curve makes it capable to detect the gauge corner cracking.

4.1.4 Guided wave testing (GWT)

GWT techniques have been widely investigated over the past decades because of the potential for long-range interrogation and detecting vertical-transverse defects under shelling and weld defects [62, 63]. They are ideal in SHM applications that can benefit from built-in transduction, moderately large inspection ranges and high sensitivity to small flaws. In rail applications, since ultrasonic guided wave can propagate through the discontinuous defects on rail surface, the screening effect of lateral cracks distributed underneath produced by surface detachment can be minimised.

Rose et al. [63] developed a GWT inspection system with non-contact air-coupled and EMATs to transmit and receive guided waves for the detection of transverse defects under shelling. Wilcox et al. [64] developed a GWT system with a dry-coupled piezoelectric transducer array to detect smooth transverse-vertical defects and alumino-thermic welds, but this system requires interruption of the operation of trains. Lanza et al. [54] developed a GWT system using a pulsed laser

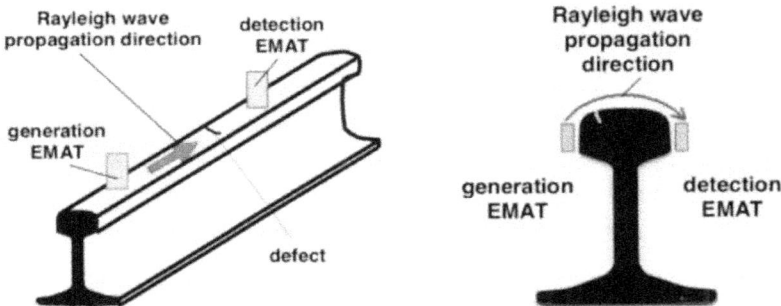

Figure 4.
Rail cracks detection using EMAT [59].

to generate ultrasonic guided waves and air-coupled transducers to sense the guided waves for the detection of vertical cracks hidden below horizontal cracks. Park et al. [65] proposed a built-in active sensing system consisting of two piezoelectric patches in conjunction with both impedance and guided wave propagation methods for rail defect detection. Marine Technology Association of South Africa and Council of Scientific and Industrial Research [66] jointly developed solar power GWT detection system (**Figure 5**). The coverage of the single system for rail defect detection is up to 2 km. Imperial College London and Guided Ultrasonics Ltd. cooperatively developed a G-shaped scanning ultrasonic rail track detection device, which can inspect vertically distributed defects and alumino-thermic weld joint [67], as shown in **Figure 6**. It can effectively inspect 18-mm-deep defects under rail crossing nose.

4.1.5 Acoustic emission testing (AET)

Different from common ultrasonic inspection, acoustic emission (AE) is instantaneous elastic waves by quick release of localised energy in solid materials under external applied loads. AE events can be captured by the piezoelectric sensors, generated elastic waves along all directions. Many sensors can be utilised to document arrival time of the signals and the variation of frequency during the crack initiation process. Hence, the nature of cracks can be determined. Through experimental study, AE has been proven a feasible solution in defecting rail detection, especially in rotating machinery [68]. A simplified analytical model, which separates defects caused by AE activities from background noise, was proposed by Thakkar et al. [48]. They also investigated the physical interaction between AE and axial load, speed, as well as traction through experiment. It is found that AE signal can be used for analysing the defects on the surface of the rail under normal operating speed.

The application of AET on rail defect detection is rarely reported. Previous research [69] shows that the benefit of AET may be limited due to imperfection of materials, which can produce different nature of signal source. Besides, the installation of the sensor may also affect the AE signal generation. It may also be affected by wheel and track defect for any misalignment [48]. Another concern of AET is signal processing for those AE waves that have similar amplitude with that of background noise produced by the rolling wheel [18]. Advanced data learning and

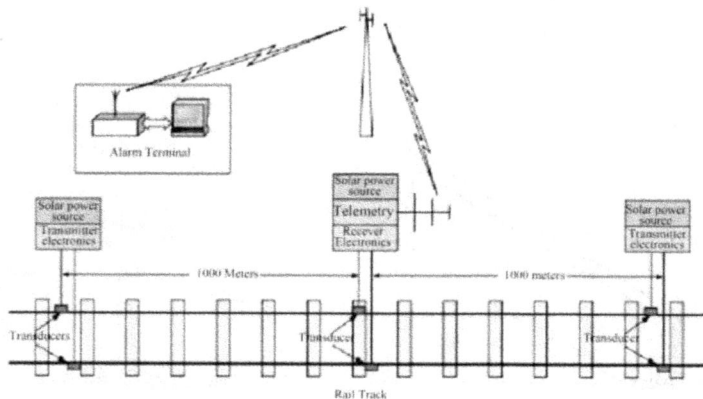

Figure 5.
Monitoring system on freight line in South Africa [66].

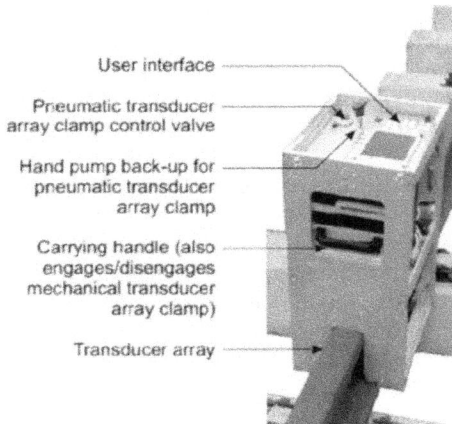

Figure 6.
G-type scan guided wave in inspection device in UK [67].

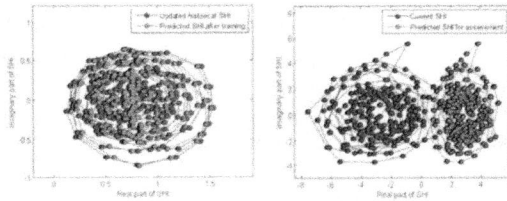

Figure 7.
(Left) Online HSR turnout condition monitoring system; (right) defect diagnosis method with model updating approach [70].

updating methods have been investigated dealing with great uncertainties arisen from the online monitoring data for more accurate and efficient damage diagnosis. AE method incorporating Bayesian framework is utilised in an online rail turnout crack monitoring system developed by CNERC-Rail [70]. The method is able to detect defects without training data of damaged rail structure and the monitoring systems have been implemented on Shanghai-Nanjing HSR lines, as shown in left panel of **Figure 7**. The rail turnout conditions are indicated in a probabilistic manner through a structure health index (SHI), as shown in right panel of **Figure 7**.

4.2 Other NDT techniques applied in rail defect detection

4.2.1 Magnetic particle testing

Magnetic particle testing [71] can be used easily to detect the specimen surface defects. But, the result is very sensitive to the specimen surface condition. If the specimen surface is coated or wet, the reliability of the detect result will decrease a lot. Therefore, removing the coating materials and surface drying are necessary before testing.

4.2.2 Eddy current testing

Eddy current testing is very simple and easy to detect surface and shallow internal cracks [28]. Eddy current sensors have been mounted on the bogie of track inspection cars and equipped in roller-guided trolleys for mobile inspection of rails [28]. They are able to detect surface and near-surface defects in the railhead but fail to locate internal defects.

4.2.3 Alternating current field measurement (ACFM)

The alternating current field measurement (ACFM) technique is an electromagnetic inspection method that uses hand-held probes, and computerised control, data acquisition and computational models. ACFM is more efficient than conventional inspection methods due to a reduced need for surface preparation and an ability to work through surface coatings. ACFM also has an added benefit that it is not only capable of detecting flaws but can also detect size defects for length and depth [72]. In 2000, TSC with the support of Bombardier Transportation began the development of an advanced ACFM system for application in the rail industry. Following the experimental work on the train axles, it became evident that an ACFM system could be deployed to detect RCF cracking on rails. This led to the development of a pedestrian-operated ACFM walking stick [73]. The inspection of the railhead is carried out by sequentially scanning across the group of sensors enabling the uninterrupted inspection of the rail. The system can detect and size gauge corner cracks and head checks smaller than 2 mm in depth. However, the ACFM sensors cannot quantify squats accurately and are unable to detect short-wave corrugation and wheelburns.

4.3 FBG-based online monitoring of HSR

In recent years, optic fibre sensors have been advocated for application to rail infrastructure monitoring. The FBG sensors have merits of being immune to electromagnetic interference (EMI) and no power supply is needed on-site. A monitoring system based on the FBG technology has been developed and installed on an operating rail line in Hong Kong for real-time and continuous detection of rail strain and temperature, rail breaks, axle counting, wheel imperfection assessment and dynamic loading identification [35]. Wang et al. [74] proposed a rail performance monitoring and safety warning system and implemented this system on a rail line by deploying FBG sensors in the rail web and at the expansion joints between supporting concrete slabs. Yoon et al. [75] proposed a distributed fibre sensory system based on Brillouin scattering and a correlation domain analysis technique for longitudinal strain monitoring of rails. Ni et al. developed a deformation monitoring system for an in-service HSR tunnel using an FBG-based monitoring system [76]. An array of FBG bending gauges was deployed at the rail slab of a segment inside the tunnel. Upon occurrences of deformation, there would be relative rotation between two adjacent bending gauges. Phase shift of the FBG sensors caused by the relative rotations was recorded, and the deformation can then be derived, resulting in a profile of the deforming rail slab, and the deformation of the tunnel can be inferred.

FBG sensors for detecting acousto-ultrasonic signals have been studied since the mid-1990s [77]. The conventional interrogation technique for FBGs as sensing elements utilises their spectral encoding and decoding capabilities for the measurand; however, the spectral decoding capability cannot be used to detect high-frequency signals (e.g., acoustic and ultrasonic waves) due to the low wavelength scanning speed. Appropriate demodulation techniques capable of high-sensitivity detection of high-frequency waves are necessary to develop acousto-ultrasonic FBG sensors. There are two main approaches to detecting acoustic and ultrasonic waves with FBGs: the first one uses a narrowband light source to

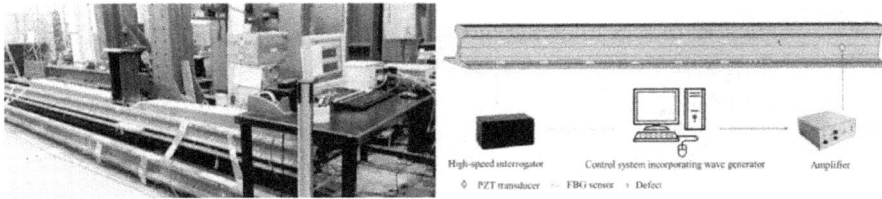

Figure 8.
Schematic set-up of the FBG-PZT monitoring system [79].

illuminate the FBGs and demodulate the power intensity variation when the waves
impinge on the FBGs, while the second one uses a broadband light source and an
optic filter. Minardo et al. [78] conducted a numerical investigation on the response
of FBGs subjected to longitudinal ultrasonic waves. Ni et al. [79] have developed a
hybrid monitoring system using FBG sensors to interrogate ultrasound signals
emitted by PZT sensors (**Figure 8**). The hybrid system has been verified in lab and a
test line in mainland China.

5. Outlooks of SHM on HSR

With embedded hybrid monitoring systems of FBG and PZT sensors, the SHM
techniques have shown their promising prospect in HSR, enabling real-time monitoring of structural conditions of in-service trains and rail infrastructure. To realise
large-scale utilisation on numerous HSR lines worldwide, practical solutions ought
to be achieved concerning both economic and efficient aspects, answering for the
need of early warning and quick decision-making upon emergencies in high-speed
operation and guiding the potential development direction of SHM applications on
HSR in the coming decades.

Wireless sensing network (WSN) provides a cost-effective approach eliminating
wires and enabling remote sensing, which largely enhances the practical applicability of SHM [80, 81]. A wireless-based system was designed to monitor the performance of rail vehicles by Nejikovsky and Keller [82]. The communication in the
WSN system can be made through satellite and Ethernet, while data are uploaded
onto cloud for storage and transmission to control room far away from site; system
data transmission plan can be found in the aforementioned railway tunnel deformation project [76]. Particularly, in terms of near-field communication, radio frequency identification (RFID) has been proposed as a competitive candidate [83],
which provides a new thinking on emerging RFID modules in normal sensors. The
passive RFID sensors embedded in the HSR structures need no wired power supply
and can be activated by passing trains, sending structural condition information.

Continuous online monitoring of HSR over multiple HSR lines puts forward the
difficulty in storage and analysis with massive data collected. The authors' team
has long been dedicated to damage diagnosis and prognosis of HSR based on monitoring data with updating and learning methods. Facing the data amount issue,
compressive sensing, which is able to sample data at sub-Nyquist sampling rate
while maintaining almost all the original information, is being actively investigated
to streamline the axle box acceleration data from an operating high-speed train
and has successfully verified the feasibility of sub-Nyquist data acquisition in HSR
online monitoring [84, 85]. This is of great significance to wireless sensing and
RFID where transmitted data amount is limited.

6. Conclusions

Various sensing technologies have long been benefiting rail industries with systematic and reliable inspection and monitoring. In turn, the vigorous development of HSR has been pushing research in sensing technologies with flourishing state-of-the-art deliverables coming out. The HSR is expanding worldwide, satisfying people's growing demands in travelling with ease and comfort, and bringing heavier inspection and maintenance tasks. In response to the expanding HSR network, conventional offline inspection will still be the primary approach taking up most of the work, and online SHM will be a powerful supporting tool playing a more important role and reflecting real-time states of the operation HSR systems. The use of sensors will be less solitary and separated but more in a combined manner containing multi-disciplinary subjects from mechanical engineering, civil engineering, electrical engineering to computer science, mathematics, etc. Moreover, the requirements to contemporary sensing go beyond fundamental functions of accuracy and reliability to flexibility, portability and environment-friendly. Taking advance of nature of railway, the SHM applications on HSR can do more than environment-friendly. The concept proposed by the authors, a high-speed train with embedded sensing systems can be treated as an integrated moving sensor, capable of gathering information not restricted to structural conditions, but air conditions inside and outside the car body concerning surrounding environment and people's health. Having accomplished multiple SHM projects on HSR lines, we are initiating just calling a start, and in the near future, the encounter of sensing technologies and HSR will continuously foster reciprocal developments, paving a high-speed path to structural well-being, sustainable environment and social health.

Acknowledgements

The authors appreciate the funding support by the Ministry of Science and Technology of China and the Innovation and Technology Commission of Hong Kong SAR Government to the Hong Kong Branch of Chinese National Rail Transit Electrification and Automation Engineering Technology Research Center (Grants Nos. 2018YFE0190100 and K-BBY1).

Author details

Lu Zhou, Xiao-Zhou Liu and Yi-Qing Ni*
National Rail Transit Electrification and Automation Engineering Technology Research Center (Hong Kong Branch), The Hong Kong Polytechnic University, Kowloon, Hong Kong

*Address all correspondence to: yiqing.ni@polyu.edu.hk

IntechOpen

References

[1] Johansson A, Nielsen JC. Out-of-round railway wheels—Wheel-rail contact forces and track response derived from field tests and numerical simulations. Proceedings of the Institution of Mechanical Engineers, Part F: Journal of Rail and Rapid Transit. 2003;**217**(2):135-146. DOI: 10.1243/095440903765762878

[2] Brunskill HP, Zhou L, Lewis R, Marshall MB, Dwyer-Joyce RS. Dynamic characterisation of the wheel-rail contact using ultrasound reflectometry. In: Proceedings of the 9th International Conference on Contact Mechanics and Wear of Rail/Wheel Systems (CM2012); Aug 27–30, 2012; Chengdu, China

[3] The Jiji Press, Ltd. Crack in Shinkansen Bullet Train Nearly Caused Derailment (News). 2017. Available from: https://www.nippon.com/en/genre/economy/l10680/ [Accessed: Dec 20, 2017]

[4] Wu Y, Du X, Zhang HJ, Wen ZF, Jin XS. Experimental analysis of the mechanism of high-order polygonal wear of wheels of a high-speed train. Journal of Zhejiang University—Science A. 2017;**18**(8):579-592. DOI: 10.1631/jzus.A1600741

[5] Nielsen J. Out-of-round railway wheels. In: Lewis R, Olofsson U, editors. Wheel-Rail Interface Handbook. 1st ed. Sawston: Woodhead Publishing; 2009. pp. 245-279. DOI: 10.1533/9781845696788.1.245

[6] Vernersson T, Caprioli S, Kabo E, Hansson H, Ekberg A. Wheel tread damage: A numerical study of railway wheel tread plasticity under thermomechanical loading. Proceedings of the Institution of Mechanical Engineers, Part F: Journal of Rail and Rapid Transit. 2010;**224**(5):435-443. DOI: 10.1243/09544097JRRT358

[7] Jin XS. Key problems faced in high-speed train operation. Journal of Zhejiang University—Science A. 2014;**15**(12):936-945. DOI: 10.1007/978-981-10-5610-9_2

[8] Kouroussis G, Connolly DP, Verlinden O. Railway-induced ground vibrations-a review of vehicle effects. International Journal of Rail Transportation. 2014;**2**(2):69-110. DOI: 10.1080/23248378.2014.897791

[9] Ikeuchi K, Handa K, Lundén R, Vernersson T. Wheel tread profile evolution for combined block braking and wheel-rail contact: Results from dynamometer experiments. Wear. 2016;**366**:310-315. DOI: 10.1016/j.wear.2016.07.004

[10] Barke DW, Chiu WK. A review of the effects of out-of-round wheels on track and vehicle components. Proceedings of the Institution of Mechanical Engineers, Part F: Journal of Rail and Rapid Transit. 2005;**219**(3):151-175. DOI: 10.1243/095440905X8853

[11] Wu TX, Thompson DJ. A hybrid model for the noise generation due to railway wheel flats. Journal of Sound and Vibration. 2002;**251**(1):115-139. DOI: 10.1006/jsvi.2001.3980

[12] Verheijen E. A survey on roughness measurements. Journal of Sound and Vibration. 2006;**293**(3):784-794. DOI: 10.1016/j.jsv.2005.08.047

[13] Wu X, Chi M. Study on stress states of a wheelset axle due to a defective wheel. Journal of Mechanical Science and Technology. 2016;**30**(11):4845-4857. DOI: 10.1007/s12206-016-1003-y

[14] Jin X, Wu L, Fang J, Zhong S, Ling L. An investigation into the mechanism of the polygonal wear of metro train

wheels and its effect on the dynamic behaviour of a wheel/rail system. Vehicle System Dynamics. 2012;**50**(12): 1817-1834. DOI: 10.1080/00423114. 2012.695022

[15] Jeffery BD, Peterson M. Assessment of rail flaw inspection data. In: AIP Conference Proceedings: Review of Progress in Quantitative Nondestructive Evaluation. United States: Springer Science; Vol. 509, No. 19. 2000. pp. 789-796. DOI: 10.1063/1.1306127

[16] Toliyat HA, Abbaszadeh K, Rahimian MM, Olson LE. Rail defect diagnosis using wavelet packet decomposition. IEEE Transactions on Industry Applications. 2003;**39**(5): 1454-1461. DOI: 10.1109/ TIA.2003.816474

[17] Magel EE. Rolling Contact Fatigue: A Comprehensive Review. Final Report. Washington, DC: Federal Railroad Administration; 2011

[18] Yilmazer P. Structural Health Condition Monitoring of Rails using Acoustic Emission Techniques. Birmingham: University of Birmingham Research; 2012

[19] Zerbst U, Lundénb R, Edel K-O, Smith RA. Introduction to the damage tolerance behaviour of railway rails—A review. Engineering Fracture Mechanics. 2009;**76**(17):2563-2601. DOI: 10.1016/j.engfracmech.2009. 09.003

[20] Li ZL, Dollevoet R, Molodova M, Zhao X. Squat growth—Some observations and the validation of numerical predictions. Wear. 2011;**271**: 148-157. DOI: 10.1016/j.wear.2010.10.051

[21] Grassie S, Nilsson P, Bjurstrom K, Frick A, Hansson LG. Alleviation of rolling contact fatigue on Sweden's heavy haul railway. Wear. 2002;**253**:42-53. DOI: 10.1016/S0043-1648(02)00081-9

[22] Müller S. A linear wheel-rail model to investigate stability and corrugation on straight track. Wear. 2000;**243**: 122-132. DOI: 10.1016/S0043-1648(00) 00434-8

[23] Alemi A, Corman F, Lodewijks G. Condition monitoring approaches for the detection of railway wheel defects. Proceedings of the Institution of Mechanical Engineers, Part F: Journal of Rail and Rapid Transit. 2017;**231**(8): 961-981. DOI: 10.1177/0954409 716656218

[24] Wang L, Xu H, Yuan H, Zhao W, Chen X. Optimizing the re-profiling strategy of metro wheels based on a data-driven wear model. European Journal of Operational Research. 2015; **242**(3):975-986. DOI: 10.1016/j. ejor.2014.10.033

[25] Salzburger HJ, Schuppmann M, Li W, Xiaorong G. In-motion ultrasonic testing of the tread of high-speed railway wheels using the inspection system AUROPA III. Insight-Non-Destructive Testing and Condition Monitoring. 2009;**51**(7):370-372. DOI: 10.1784/insi.2009.51.7.370

[26] Cavuto A, Martarelli M, Pandarese G, Revel GM, Tomasini EP. Train wheel diagnostics by laser ultrasonics. Measurement. 2016;**20180**:99-107. DOI: 10.1016/j.measurement.2015.11.014

[27] Kenderian SHANT, Djordjevic BB, Cerniglia D, Garcia G. Dynamic railroad inspection using the laser-air hybrid ultrasonic technique. Insight-Non-Destructive Testing and Condition Monitoring. 2006;**48**(6):336-341. DOI: 10.1784/insi.2006.48.6.336

[28] Pohl R, Erhard A, Montag HJ, Thomas HM, Wüstenberg H. NDT techniques for railroad wheel and gauge corner inspection. NDT&E International. 2004;**37**(2):89-94. DOI: 10.1016/j.ndteint.2003.06.001

[29] Hwang J, Lee J, Kwon S. The application of a differential-type Hall sensors array to the nondestructive testing of express train wheels. NDT&E International. 2009;**42**(1):34-41. DOI: 10.1016/j.ndteint.2008.08.001

[30] Brizuela J, Fritsch C, Ibanez A. Railway wheel-flat detection and measurement by ultrasound. Transportation Research Part C: Emerging Technologies. 2011;**19**(6): 975-984. DOI: 10.1016/j.trc.2011.04.004

[31] Brizuela J, Fritsch C, Ibáñez A. New ultrasonic techniques for detecting and quantifying railway wheel-flats. In: Reliability and Safety in Railway. London: InTech; 2012. pp. 399-418. DOI: 10.5772/35236

[32] Nielsen JC, Oscarsson J. Simulation of dynamic train-track interaction with state-dependent track properties. Journal of Sound and Vibration. 2004; **275**(3):515-532. DOI: 10.1016/j. jsv.2003.06.033

[33] Stratman B, Liu Y, Mahadevan S. Structural health monitoring of railroad wheels using wheel impact load detectors. Journal of Failure Analysis and Prevention. 2007;7(3):218-225. DOI: 10.1007/s11668-007-9043-3

[34] Filograno ML, Corredera P, Rodríguez-Plaza M, Andrés-Alguacil A, González-Herráez M. Wheel flat detection in high-speed railway systems using fiber Bragg gratings. IEEE Sensors Journal. 2013;**13**(12):4808-4816. DOI: 10.1109/JSEN.2013.2274008

[35] Tam HY, Lee T, Ho SL, Haber T, Graver T, Méndez A. Utilization of fiber optic Bragg Grating sensing systems for health monitoring in railway applications. In: Proceedings of the 6th International Workshop on Structural Health Monitoring. Vol. 2007. Stanford, CA: DEStech Publications, Inc.; Sep 11–13, 2007. pp. 1889-1896

[36] Liu XZ, Ni YQ. Wheel tread defect detection for high-speed trains using FBG-based online monitoring techniques. Smart Structures and Systems. 2018;**21**(5):687-694. DOI: 10.12989/sss.2018.21.5.687

[37] Skarlatos D, Karakasis K, Trochidis A. Railway wheel fault diagnosis using a fuzzy-logic method. Applied Acoustics. 2004;**65**(10):951-966. DOI: 10.1016/j. apacoust.2004.04.003

[38] Belotti V, Crenna F, Michelini RC, Rossi GB. Wheel-flat diagnostic tool via wavelet transform. Mechanical Systems and Signal Processing. 2006;**20**(8): 1953-1966. DOI: 10.1016/j. ymssp.2005.12.012

[39] Seco M, Sanchez E, Vinolas J. Monitoring wheel defects on a metro line: System description, analysis and results. WIT Transactions on The Built Environment. 2006;**88**:973-982. DOI: 10.2495/CR060951

[40] Amini A. Online condition monitoring of railway wheelsets [Doctoral dissertation]. Birmingham: University of Birmingham; 2016

[41] Lu S, He Q, Hu F, Kong F. Sequential multiscale noise tuning stochastic resonance for train bearing fault diagnosis in an embedded system. IEEE Transactions on Instrumentation and Measurement. 2014;**63**(1):106-116. DOI: 10.1109/TIM.2013.2275241

[42] Wang J, He Q, Kong F. A new synthetic detection technique for trackside acoustic identification of railroad roller bearing defects. Applied Acoustics. 2014;**85**:69-81. DOI: 10.1016/ j.apacoust.2014.04.005

[43] Lagnebäck R. Evaluation of wayside condition monitoring technologies for condition-based maintenance of railway vehicles [Doctoral dissertation]. Luleå: Luleå tekniska universitet; 2007

[44] Asplund M, Lin J. Evaluating the measurement capability of a wheel profile measurement system by using GR&R. Measurement. 2016;**92**:19-27. DOI: 10.1016/j.Measurement.2016.05.090

[45] Zhang ZF, Gao Z, Liu YY, Jiang FC, Yang YL, Ren YF, et al. Computer vision based method and system for online measurement of geometric parameters of train wheel sets. Sensors. 2011;**12**(1):334-346. DOI: 10.3390/s120100334

[46] Chen Y, Xing Z, Li Y, Yang Z. Standard-wheel-based field calibration method for railway wheelset diameter online measuring system. Applied Optics. 2017;**56**(10):2714-2723. DOI: 10.1364/AO.56.002714

[47] Wang J, Jian Z, Li Y, Masanori S. Geometric parameters measurement of wheel tread based on line structured light. Journal of Robotics, Networking and Artificial Life. 2016;**3**(2):124-127. DOI: 10.2991/jrnal.2016.3.2.12

[48] Thakkar NA, Steel JA, Reuben RL. Rail–wheel interaction monitoring using acoustic emission: A laboratory study of normal rolling signals with natural rail defects. Mechanical Systems and Signal Processing. 2010;**24**(1):256-266. DOI: 10.1016/j.ymssp.2009.06.007

[49] Clark R. Rail flaw detection: Overview and needs for future developments. NDT&E International. 2004;**37**(2):111-118. DOI: 10.1016/j.ndteint.2003.06.002

[50] Papaelias MP, Roberts C, Davis CL. A review on non-destructive evaluation of rails: State-of-the-art and future development. Proceedings of the Institution of Mechanical Engineers, Part F: Journal of Rail and Rapid Transit. 2008;**222**(4):367-384. DOI: 10.1243/09544097JRRT209

[51] Lanza SF. Advances in non-contact ultrasonic inspection of rail tracks.

Experimental Techniques. 2000;**24**:23-26. DOI: 10.1111/j.1747-1567.2000.tb02352.x

[52] Nielsen SA, Bardenshtein AL, Thommesen AM, Stenum B. Automatic laser ultrasonics for rail inspection. In: Proceedings of the 16th World Conference on Non-Destructive Testing. NDT.net; Montreal, Canada; 2004

[53] Kenderian S, Djordjevic BB, Cerniglia D, Garcia G. Laser-air hybrid ultrasonic technique for dynamic railroad inspection applications. Insight-Non-Destructive Testing and Condition Monitoring. 2005;**47**(6):366-340. DOI: 0.1784/insi.47.6.336.66454

[54] Lanza SF, Rizzo P, Coccia S, Bartoli I, Fateh M, Viola E, et al. Non-contact ultrasonic inspection of rails and signal processing for automatic defect detection and classification. Insight. 2005;**47**:1-8. DOI: 10.1784/insi.47.6.346.66449

[55] Utrata D, Clark R. Groundwork for rail flaw detection using ultrasonic phased array inspection. In: Thompson DO, Chimenti DE, editors. Review of Progress in Quantitative Nondestructive Evaluation. New York, NY: American Institute of Physics; 2003. p. 22

[56] Wooh SC, Wang J. Nondestructive characterization of defects using a novel hybrid ultrasonic array sensor. NDT&E International. 2002;**35**:155-163. DOI: 10.1016/S0963-8695(01)00038-X

[57] Alaix R. High speed rail testing with phased array probes. In: Proceedings of the 7th World Congress on Railway Research. Montreal, Canada; 2006

[58] Garcia G, Zhang J. Application of Ultrasonic Phased Arrays for Rail Flaw Inspection. Washington, DC: U.S. Department of Transportation, Federal Railroad Administration, Office of Research and Development; 2006

[59] Palmer SB, Dixon S, Edwards RS, Jian X. Transverse and longitudinal crack detection in the head of rail tracks using Rayleigh wave-like wideband guided ultrasonic waves. In: Proceedings of SPIE - The International Society for Optical Engineering. United States: Society of Photo-optical Instrumentation Engineers. Bellingham, WA, 2005. doi: 10.1117/12.598142

[60] Edwards RS, Sophian A, Dixon S, Tian GY, Jian X. Dual EMAT and PEC non-contact probe: Applications to defect testing. NDT&E International. 2006;39:45-52. DOI: 10.1016/j.ndteint.2005.06.001

[61] Fan Y, Dixon S, Edwards RS, Jian X. Ultrasonic surface wave propagation and interaction with surface defects on rail track head. NDT&E International. 2007;40(6):471-477. DOI: 10.1016/j.ndteint.2007.01.008

[62] Loveday PW. Guided wave inspection and monitoring of railway track. Journal of Nondestructive Evaluation. 2012;31:303-309. DOI: 10.1007/s10921-012-0145-9

[63] Rose JL, Avioli MJ, Mudge P, Sanderson R. Guided wave inspection potential of defects in rail. NDT&E International. 2004;37:153-161. DOI: 10.1016/j.ndteint.2003.04.001

[64] Wilcox P, Evans M, Pavlakovic B, Alleyne D, Vine K, Cawley P, et al. Guided wave testing of rail. Insight-Non-Destructive Testing and Condition Monitoring. 2003;45(6):413-420. DOI: 10.1784/insi.45.6.413.52892

[65] Park S, Inman DJ, Lee JJ, Yun CB. Piezoelectric sensor-based health monitoring of railroad tracks using a two-step support vector machine classifier. Journal of Infrastructure Systems, ASCE. 2008;14:80-88. DOI: 10.1061/(ASCE) 1076-0342(2008)14:1(80)

[66] Burger FA. A practical continuous operating rail break detection system using guided waves. In: Proceedings of the 18th World Conference on Non-Destructive Testing. South Africa: South African Institute for Non-Destructive Testing; Ap. 16–20, 2012; Durban, South Africa

[67] Cawley P, Wilcox P, Alleyne DN, Pavlakovic B, Evans M, Vine K. Long range inspection of rail using guided waves-field experience. Review of Progress in Quantitative Nondestructive Evaluation. 2003;22:236-243. DOI: 10.1.1.159.6462

[68] Bruzelius K, Mba D. An initial investigation on the potential applicability of acoustic emission to rail track fault detection. NDT&E International. 2004;37(7):507-516. DOI: 10.1016/j.ndteint.2004.02.001

[69] Christian UG, Ohtsu M. Acoustic Emission Testing. Berlin: Springer; 2008

[70] Wang J, Liu XZ, Ni YQ. A Bayesian probabilistic approach for acoustic emission-based rail condition assessment. Computer-Aided Civil and Infrastructure Engineering. 2018;33(1): 21-34. DOI: 10.1111/mice.12316

[71] Mix PE. Magnetic particle testing. In: Introduction to Nondestructive Testing: A Training Guide. 2nd ed. Hoboken, NJ, USA: John Wiley & Sons, Inc.; 2005. 680 p. DOI: 10.1002/0471719145.ch7

[72] LeTessier R, Coade R, Geneve B. Sizing of cracks using the alternating current field measurement technique. International Journal of Pressure Vessels and Piping. 2002;79:549-554. DOI: 10.1016/S0308-0161(02)00088-1

[73] Papaelias MP, Lugg MC, Roberts C, Davis CL. High-speed inspection of rails using ACFM techniques. NDT&E International. 2009;42:328-335. DOI: 10.1016/j.ndteint.2008.12.008

[74] Wang CY, Tsai HC, Chen CS, Wang HL. Railway track performance monitoring and safety warning system. Journal of Performance of Constructed Facilities, ASCE. 2011;**25**:577-585. DOI: 10.1061/(ASCE)CF.1943-5509.0000186

[75] Yoon HJ, Song KY, Kim JS, Kim DS. Longitudinal strain monitoring of rail using a distributed fiber sensor based on Brillouin optical correlation domain analysis. NDT&E International. 2011; **44**:637-644. DOI: 10.1016/j.ndteint. 2011.07.004

[76] Zhou L, Zhang C, Ni YQ, Wang CY. Real-time condition assessment of railway tunnel deformation using an FBG-based monitoring system. Smart Structures & Systems. 2018;**21**(5): 537-548. DOI: 10.12989/sss.2018. 21.5.000

[77] Wild G, Hinckley S. Acousto-ultrasonic optical fiber sensors: Overview and state-of-the-art. IEEE Sensors Journal. 2008;**8**:1184-1193. DOI: 10.1109/JSEN.2008.926894

[78] Minardo A, Cusano A, Bernini R, Zeni L, Giordano M. Response of fiber Bragg gratings to longitudinal ultrasonic waves. IEEE Transactions on Ultrasonics, Ferroelectrics, and Frequency Control. 2005;**52**:304-312. DOI: 10.1109/TUFFC.2005.1406556

[79] Wang JF, Yuan MD, Ni YQ. Rail crack monitoring using fiber optic based ultrasonic guided wave detection technology. In: Proceedings of the 11th International Workshop on Structural Health Monitoring 2017. United States: DEStech Publications, Inc.; Sep 12–14, 2017, Stanford, California, USA

[80] César BR, Ke G, Yin XF, Thomas K. Wireless communications in smart rail transportation systems. Wireless Communications and Mobile Computing. 2017. 10 pages. DOI: 10.1155/2017/6802027

[81] Ruscelli AL, Cecchetti G, Sgambelluri A, Cuginiy F, Giorgetti A, Paolucci F, et al. Wireless communications in railway systems. In: The Seventh International Conference on Mobile Services, Resources, and Users. New York: Association for Computing Machinery; Jun 25–29, 2017; Venice, Italy

[82] Nejikovsky B, Keller E. Wireless communications based system to monitor performance of rail vehicles. In: Proceedings 2000 ASME/IEEE Joint Railroad Conference. United States: IEEE; Ap. 6, 2000; Newark, NJ, United States

[83] Ravdeep K, Ramin K, Parida A, Uday K. Applications of radio frequency identification (RFID) technology with eMaintenance cloud for railway system. International Journal of System Assurance Engineering and Management. 2013;**5**:99-106. DOI: 10.1007/s13198-013-0196-z

[84] Ni YQ, Chen SX. Compressive sensing for vibration signals in high-speed rail monitoring. In: Proceedings of the 9th European Workshop on Structural Health Monitoring. NDT.net; Jul 10–13, 2018; Manchester, UK

[85] Chen SX, Ni YQ. Compressive sensing for high-speed rail condition monitoring using redundant dictionary and joint reconstruction. In: Proceedings of the 9th European Workshop on Structural Health Monitoring. NDT.net; Jul 10–13, 2018; Manchester, UK

Chapter 4

Main Ways to Improve Cutting Tools for Machine Wheel Tread Profile

Alexey Vereschaka, Popov Alexey, Grigoriev Sergey, Kulikov Mikhail and Sotova Catherine

Abstract

This chapter considers the methods to increase the performance and reliability of the reprofile machining of the wheel tread profile. Proceeding from the fact that both in milling and turning, the cutting tool is a key element to ensure performance and reliability of the manufacturing process, the study considers the methods to increase the performance properties of cutting tools. In particular, the study includes the investigation of the following ways to improve cutting tools (carbide inserts) to machine wheel tread profile: replacement of traditional grades of WC-TiC-Co carbides with more efficient ones based on WC-TiC-TaC-Co; application of special thermally conductive pads, gaskets, and pastes to improve the distribution of heat flows in the cutting zone; and application of modern nanoscale composite multilayer coatings (NMCC). It is noted that even higher performance can be obtained by combining the above three methods, in particular, by combining application of special thermal pads and NMCC.

Keywords: reprofile machining, wheel tread profile, thermally conductive pads, nanoscale composite multilayer coatings

1. Introduction

In connection with the change in the profile geometry of the wheel tread due to mechanical wear and plastic deformations, as well as appearance of thermomechanical damages, which result from braking of the rolling stock and which occur as flat sections of cold hardening (slides) with the formation of local martensitic structures with hardness of up to 850–900 HV (white spots), there is a need for a regular reprofiling of the wheel treads. Considering the fact that Russia possesses one of the longest rail networks in the world (the second only to the USA), it is clear that the specified challenge is very comprehensive (given more than 6 million wheel sets in operation). There are two main methods of reprofiling: milling and turning. At the same time, the performance and tool life indicators of cutting tools (carbide inserts) are an important and sometimes a critical factor. Milling the profile of wheel treads is the most common process applied to machine wheel sets for locomotives and electric trains.

The existing technology for machining wheel sets on KZH-20 machines is one of the most complex and time-consuming machining operations performed in depot conditions during maintenance and current repairs of locomotives and multiple-unit and subway rolling stocks. KZH-20 is a wheel-milling machine to mill wheel sets without rolling them out from under a locomotive, manufactured at the KZTZ (Kramatorsk Heavy Duty Machine Tool Building Plant, Ukraine, see **Figure 1a**). A similar pattern to restore the profile of wheels with a special shaped milling cutter with replaceable carbide inserts, except for machines of KZH-20 type, is also implemented on machines of Simmons Machine Tool Corporation (USA) and Kawasaki Inc. (Japan). The main advantages of the above method, in comparison with turning, are as follows:

- a possibility of machining with minimum depths of cut in a single pass, which makes it possible to increase the number of possible resharpening of the wheel tread profile and, consequently, to increase the life of wheel sets;

- an increase in the cutting speed and, consequently, an increase in the machining capacity, provided by increasing the rotational speed of the milling cutter and not the wheel set;

- an increase in the tool life of the cutting tool by increasing the active length of the cutting edge in simultaneous cutting with several cutter teeth and reducing the overall heat stress of the cutting process;

- enhancement of the safety of machine tool operators due to safer and more transportable shape of chips in milling.

Furthermore, it was found that milling, in contrast to turning, provides practically unimpeded mechanical machining of the wheel tread profile of reinforced tires and tires with increased hardness.

Despite the above advantages, the most widespread pattern for wheel milling with shaped cylindrical cutters (see **Figure 1b**) has the following significant drawbacks:

- poor efficiency in comparison with turning (the main machining time for a single wheel set on a machine of KZH-20 type is 0.6–0.8 h on average);

- roughness of the machined surface and the accuracy of machining the wheel tread profile are related to the accuracy and quality of manufacturing and maintaining shaped milling cutters;

- high complexity of manufacturing and labor-intensive maintenance of special shaped milling cutters (auxiliary time for tool maintenance after machining each wheel set can reach 30–45 min);

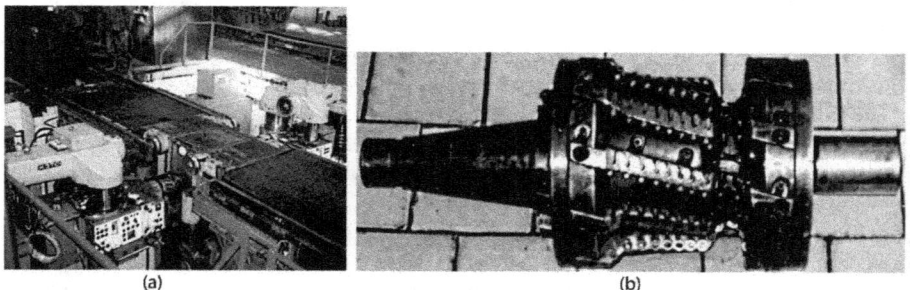

(a)   (b)

Figure 1.
Process of machining wheel treads on KZH-20 machines (a) and a milling cutter for wheel tread machining (b).

- need to purchase a large number of fairly expensive carbide inserts (123–125 inserts per milling cutter) with substantial consumption; and

- possibility of machining wheel sets under single profile provided by shaped cutting tools (regardless of the wheel wear degree).

The on-the-site studies found that to date, locomotive depots use various replaceable and brazed carbide inserts. The diagram to show ranges for shapes of carbide inserts used in locomotive depots in different regions of Russia is presented in **Figure 2** [1].

As can be seen from the diagram, ISO RNGN 121200 carbide inserts used in shaped milling cutters in machines of KZH-20 type are the most widely used cutting tools for mechanical machining of the wheel tread profile. With the actual distribution of average annual consumption in absolute quantitative terms, i.e., in the weight of the consumable carbide, inserts of the RNGN 121200 type gain 53.9% of the weight of the carbide consumed at locomotive and multiple-unit rolling stock repair enterprises of the JSC Russian Railways. Meanwhile, inserts of the LNMX type are also widely used for turning the railway wheel sets with rolling out from under the rolling stock on wheel-turning machines RAFAMET S.A. (Poland) Model UBB-112, and Hegenscheidt-MFD GmbH (Germany) (Model U2000-400) [1].

The process of manufacturing critical parts of railway rolling stock (turning of wheel sets, boring of wheel bands, turning of axes, etc.) is accompanied by:

- high removal of withdrawn allowance at cutting depth a_p = 5–20 mm, feed f = 0.8–1.5 mm/rev, and cutting speed v_c = 30–50 m/min,

- high variation of cutting allowance (radial runout may reach up to 15 mm),

- inclusion of nonmetallic particles with increased abrasive properties on the machined surface of forged slab [2–4].

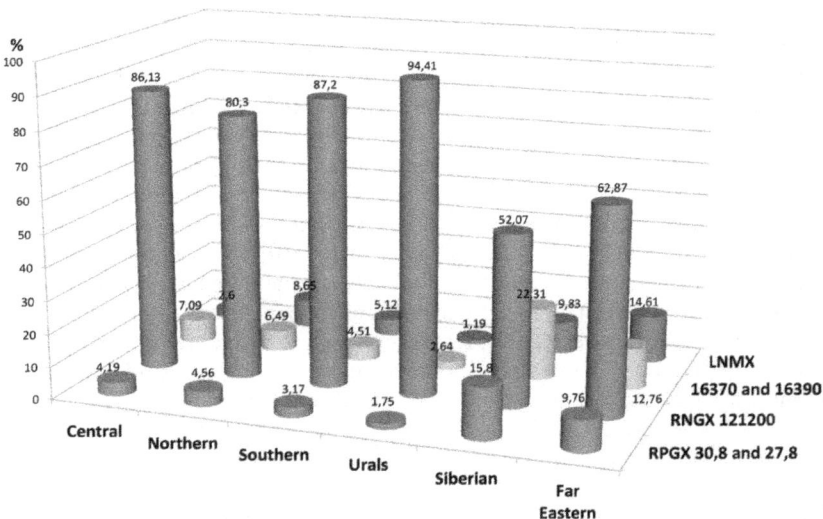

Figure 2.
Percentage distribution of shapes of carbide inserts used in locomotive depots in different regions of Russia [1].

Edge machining of workpieces under the above conditions produces elevated heating of the cutting area (up to 800–1000°C), which results in high concentration of thermal stresses directly at the contact areas of carbide inserts (for example, inserts of LNMX ISO shape) used in this process of manufacturing products for rail transport [2–7]. The studies of wear mechanisms of cutting carbide inserts with coatings of various compositions have shown [3, 9–16] that the process of wear of inserts under conditions of the high thermal stresses is accompanied by thermo-plastic deformation of a cutting edge. This in turn is connected with the subsequent intense failure of coating and high adhesion and fatigue wear, which is accompanied by chipping of cutting edges or complete failure of fragile cutting part of a tool [3, 9–11, 14–16].

In this regard, the decrease in thermal stress of the cutting area by the deposition of nanoscale multilayer composite coatings (NMCCs) on the working surfaces of the tool, which reduce friction and capacity of heat sources, as well as the general improvement of the conditions of heat transfer out of the cutting area improves the tool life and the efficiency of the HPC processes. The studies of the effect of wear-resistant coatings on the thermal state of the cutting system under severe cutting conditions [11–13] have shown that they reduce thermal and mechanical loads on the tool and increase its efficiency.

The standard method for reducing thermal stress of the cutting process includes the use of cutting fluids. However, under heavy conditions of machining, the efficiency of cutting fluids decreases significantly. Besides, specialized machine equipment (including wheel turning machines and vertical turning machines), intended for manufacturing of products (wheel sets, wheel bands, axles, etc.) used in rail transport, does not use the systems of supply of liquid fluids because of high probability of their intense damage. Thus, the main objective of this study was to develop a tool system improving the efficiency of the technology of heavy machining of workpieces of rail rolling stock products by reducing the thermal stress of the cutting process and cutting tools.

2. Effect of carbide grade on machining parameters

The study investigated the inserts made of tool carbides of (WC-TiC-Co) and (WC-TiC-TaC-Co) groups (without wear-resistant coatings). The machining was conducted at standard cutting modes typical for locomotive depots: the milling cutter rotational speed was 93 rpm (at the cutting speed of 60 m/min), the working feed rate was 120–160 mm/min, and the allowance was taken in 23 passes (at the depth of cut of 23 mm per pass). Complete failure (blunting) of more permissible value observed on the tested carbide inserts due to macro- and microchipping along the periphery of the rake face was taken as a criterion for blunting of the tested carbide inserts. The tool life indicators for the tested carbide inserts were defined through the ratio between the number of wheel sets machined and a complex set of inserts made of a certain grade of carbide (considering normally worn, broken, inverted, and replaced inserts). The test results are presented in **Figure 3**.

The analysis of the obtained results shows that the modern carbides of the (WC-TiC-TaC-Co) group have wear resistance 1.5–2.0 times higher than the carbides of the (WC-TiC-Co) group. Meanwhile, the change in manufacturing conditions produces less effect on wear resistance of the carbides of the (WC-TiC-TaC-Co) group.

Following the analysis of the external wear pattern and the structure of failures of the cutting edges on the standard RNGX 121200-shaped carbide inserts, it can be noted that in milling tire steel, for the carbide inserts of the (WC-TiC-Co)

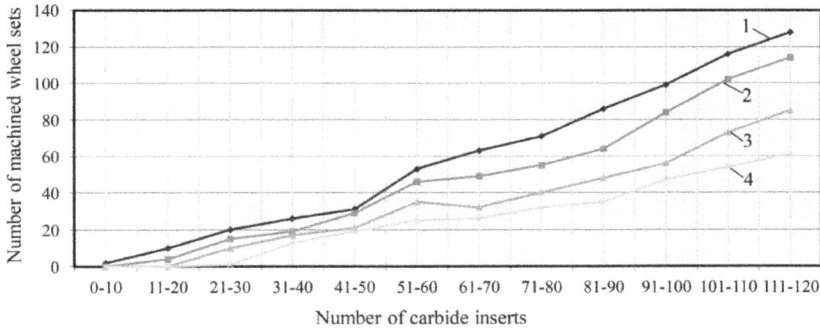

Figure 3.
Results for the tests of RNGX 121200-type carbide inserts made of various carbide grades: 1—WC-15%TiC-6%Co, 2—WC-14%TiC-8%Co, 3—T5 (WC-TiC-TaC-Co), and 4—T1 (WC-TiC-TaC-Co).

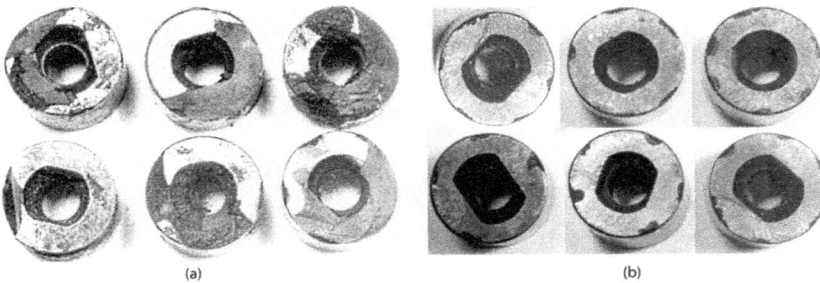

Figure 4.
Failure patterns on (WC-TiC-Co) (a) and (WC-TiC-TaC-Co) carbide inserts (b).

(78.0%WC, 14.0%TiC, 8.0%Co and 79.0%WC, 15.0%TiC, 6.0%Co) grades, currently most widely used for wheel milling, the most typical mechanism of failure is brittle fracture of the cutting edges as macrochipping on the rake and flank faces of the carbide inserts with depth over 3 mm (see **Figure 4a**). Meanwhile, for cutting inserts of the (WC-TiC-TaC-Co) carbides (T1 grade 79%WC, 4.4%TiC, 3.6%TaC, 5.8%Co and T5 grade 78.2%WC, 4.0%TiC, 5.0%TaC, 6.0%Co), failure occurs as blunting of the cutting edges due to wear and microchipping along the periphery of the rake face, due to contact-fatigue chipping with an area of 1.5–2.0 mm^2 and a depth of 0.3–0.5 mm (see **Figure 4b**). The obtained results are in good agreement with the data of production tests of various carbide grades in wheel milling of wheel sets of locomotives, electric, and subway trains on machines of the KZH-20 type [1].

The investigation of the wear rate of the carbide inserts of various carbide grades along the wheel tread profile is shown in **Figure 5**. The analysis of **Figure 4** shows that for the (WC-TiC-Co) and (WC-TiC-TaC-Co) carbides groups, the highest wear rate of the carbide inserts was registered on the area of the wheel tread until the circular cutting of ridge (with the maximum in a plane from the wheel rolling circle), with the presence of thermomechanical defects on the wheel tread (slides, white spots, chipping, etc.).

A slight increase in the wear rate was also registered at the ridge top where pointed rolling appeared. Meanwhile, it should be noted that the wear rate of carbide inserts of the (WC-TiC-TaC-Co) carbide group is more uniform in the wheel tread profile, which makes it possible to significantly reduce the tool costs by installing carbide inserts of various carbide grades in the milling cutters. Thus, the fifth-tenth cutter knife seats on the wheel rolling circle face should

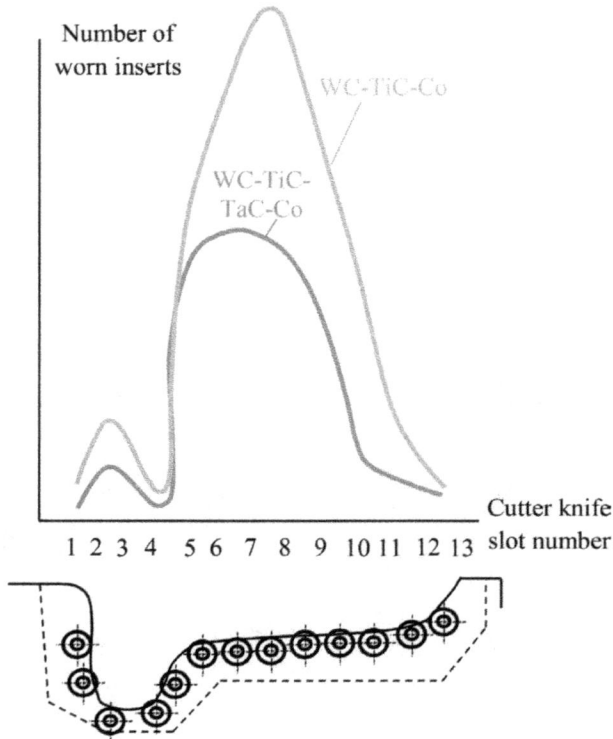

Figure 5.
Wear rates for the carbide inserts of various carbide grades depending on their placement in cutter knife seats.

bear more stable expensive carbides of the (WC-TiC-TaC-Co) group, while cheaper carbides of the (WC-TiC-Co) group can be installed to less critical sections of the profile. This technology provides a decrease in the cutting tool costs by 30% at maximum.

According to the basic principles of the theory of metal cutting, in turning and milling with carbide inserts, the changes in the pattern of the chip formation process and the temperature-stressed state on the tool cutting edges are determined by the geometric parameters of the sharpening of the cutting edges on the carbide inserts. At present, for the modern carbide inserts used to machine wheel sets of locomotives (LNMX 191940, SNMM 190616, RPUX 2709MO, and RNGX 121200 shapes), the geometry of the cutting edge sharpening is determined by two main parameters: the inclination angle of the negative reinforcing wear land (γ_F) on the rake face and the width of the negative reinforcing wear land (f).

In [17, 18], it was found that under conditions of intermittent cutting, the creation of a negative wear land on the rake face of the carbide insert increases the mechanical and heat resistance of the cutting edge. If there is a negative wear land on the cutting edge, the center of chip pressure shifts from the top of the cutting wedge and thereby increases the cutting edge strength. In this case, the angle of sharpening exceeds 90°, and the cutting wedge starts working under the conditions of compression deformation (in contrast to the bending conditions when a sharp tool is used). Moreover, an increase in the angle of sharpening improves the conditions for the heat transfer from the cutting edge to the tool body.

For carbide inserts used for wheel turning of locomotive tires (LNMX 191940, SNMM 190616, and RPUX 2709MO shapes), the maximum wear land width is limited by the start of the chip-breaking groove and is f = 0.4–0.6 mm on average. The carbide inserts of the RNGX 121200 shape (used for wheel milling) have the reinforcing wear land of f = 0.1–0.2 mm on average. The special study to determine the effect of the width of the reinforcing wear land f on the tool life of cutting tools in machining the tire steel showed (see **Figure 6**) that regardless of the tool material properties, a reduction in the wear land width led to a significant decline in the tool life indicator. For example, for inserts made of the (WC-TiC-TaC-Co) carbides with high hardness of carbide matrix and the (WC-TiC-Co) carbides with high brittleness of carbide matrix, a reduction of the reinforcing wear land by two times led to a decrease in the tool life indicators for the inserts on average by 45–55% and 80–90%, respectively. The analysis of the wear patterns on the inserts showed the presence of considerable plastic deformation of the cutting edge, while for the (WC-TiC-Co) carbides, in the area of maximum wear of the carbide insert cutting edge, there were formation centers of microcracks, chipping, and brittle fracture.

According to the data of [18], finishing-reinforcing machining by the method of cutting edge rounding makes it possible to achieve an increase in the depth of penetration of residual compressive stresses with simultaneous reduction of their gradient in a thin surface layer; i.e., it leads to the creation of a favorable profile of the residual compressive stresses in the near-surface layer of the insert. Meanwhile, during cutting, a tip of the cutting wedge will be under the action of compressive rather than tensile stresses, like in cutting with a sharp cutting edge, and smooth rounding of the edges ensures no voltage concentrators [17]. The production experimental tests conducted in wheel milling of wheel sets of locomotives and subway trains on machines of the KZH-20 type showed that the use of the inserts of the RNGX 121200 shape from the (WC-TiC-Co) and (WC-TiC-TaC-Co) carbide groups with a radius of rounding r = 0.06–0.08 mm increases the tool life indicators by 30 and 25%, respectively, with a general reduction in the number of large chipping and chipping of the cutting edges, as compared to the inserts with the width of the sharpened negative wear land of 0.1–0.2 mm.

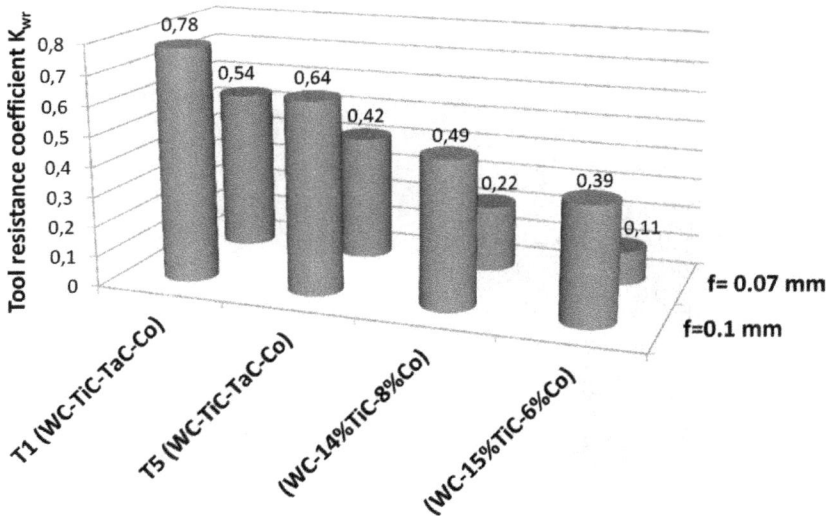

Figure 6.
Relation between the width of the reinforcing wear land f and the tool resistance of the cutting tools made of various carbide grades in machining the tire steel, where K_{wr} is tool resistance coefficient, at f = 0.15 mm, K_{wr} = 1.0.

3. Application of thermal conductive paste and pads in combination with modifying coatings

Along with the inserts of the RNGX 121200 shape, the inserts of the LNMX 301940 shape are also actively used.

The study of the process of wearing of cutting tool equipped with carbide inserts in machining wheel pair contour showed that wearing is accompanied with ductile deformation of cutting wedge of carbide tool followed by brittle fracture (**Figure 7**).

The study mainly focused on testing carbide inserts, which were mounted in tool holders of the cutting tool assemblies. The selection of the shapes of two-way inserts was justified by the extensive use of such inserts in machining rail rolling stock products. Because of large rake angles (γ = 12–15°) and wide chip-breaking grooves, at rake face of carbide inserts (width 2.5–3.5 mm), large air cavities are formed in the contact area of bearing surfaces of the inserts and the tool holder, and the total area of their actual contact can reach up to 50–65% of the total contact area. The above fact results in significant deterioration of heat transfer from carbide insert to holder body, which is a massive heat absorber, since the air thermal conductivity is 3000 times lower than the thermal conductivity of metal of the tool holder (**Figure 8**). With this in mind, during the development of a tool system with improved heat transfer from carbide insert to bearing surface of tool holder, elastic pads of ceramic-polymeric sheet reinforced material with high thermal conductivity were mounted on. In shape and thickness, the above elastic pads corresponded to the sizes of the chip-breaking grooves of carbide inserts [8]. The used ceramic-polymer pads are characterized by high elasticity (at least 50%) and thermal conductivity of about 0.8–1.4 W/(m K), which provided a significant increase in heat transfer along the entire bearing surface of the insert by reducing air gaps between bearing surfaces of the carbide insert and the tool holder (**Figure 8b**). Due to fiber glass reinforcement, ceramic-polymer pads withstand compression of up to 40 MPa, and that guarantees reliable mounting of carbide insert. With the change of bearing surface of carbide insert, when the previous bearing surface of the insert

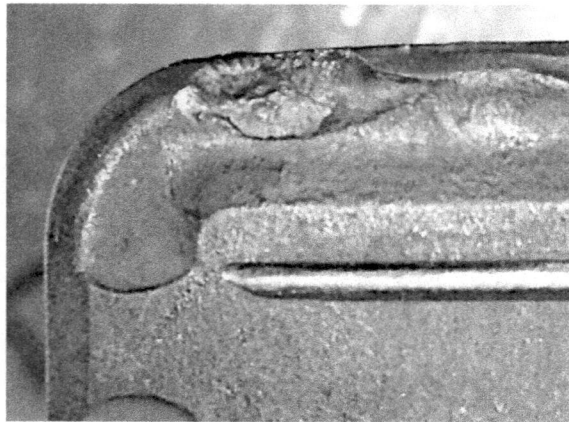

Figure 7.
Chipping on the cutting insert of LNMX 301940 shape on the rake face of carbide (14% TiC, 78% WC, 8% Co) with nanostructured multilayered composite coating (NMCC) Ti-TiN-TiAlN in machining of running surface of wheel pair with a_p = 6.0 mm, f = 1.2 mm, and v_c = 70 m/min.

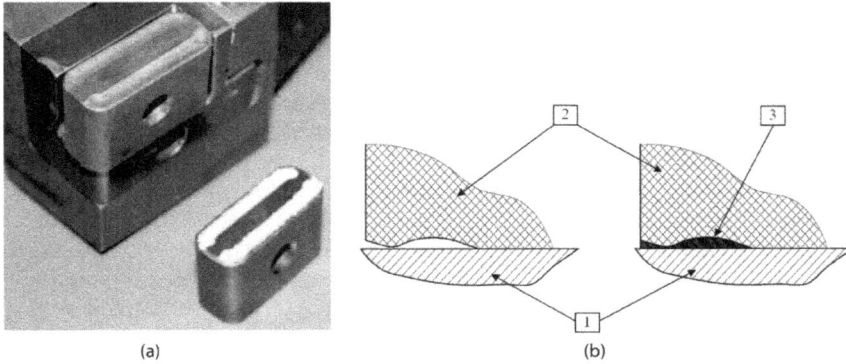

Figure 8.
(a) General view of the insert, and (b) contact of bearing surfaces of holder and cutting insert: 1—holder, 2—carbide insert, and 3—thermal pad.

became the rake face, the remains of the pads appearing in the cutting area were easily removed by chips cutoff.

4. Development of the system of increasing thermal conductivity of cemented carbide

It is known that the contact area in all friction pairs is determined not by the nominal and by the actual contact area, which is the total area of the contacting asperities of microroughness of friction pair and which comprises some percentages of the nominal contact area [8]. The air areas (pockets) are formed in places with absence of contact. The thermal conductivity of air is 3500 times lower than that of metals used in composite cutters. Therefore, the border between the contact surfaces of cutting insert and tool holder has high thermal resistance, and that greatly deteriorates the conditions of heat transfer into the tool.

Reduction of the thermal resistance in the narrow area of contact supporting surfaces is proposed by use of thermally-conductive interface with increased thermal conductivity. The specified interface formed by special thermally-conductive paste which thin layer is located between the contacting surfaces of tool holder and carbide cutting insert (**Figure 9**). Paste AlSink-3 on the basis of a mixture of oxides of aluminum and zinc together with a silicon organic solvent.

To ensure normal heat transfer, all air should be eliminated from gaps by special elastic thermally conductive composition with much higher thermal conductivity. However, in any case, the thermal properties of the best thermally conductive

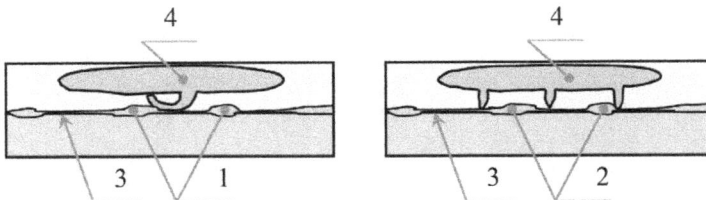

Figure 9.
Example of supporting surfaces of the cutting insert and tool holder [2]: 1—air gap clearance, 2—thermally conductive paste, 3—supporting surfaces of the carbide inserts and tool holder, and 4—heat flow.

pastes are lower than that of metals, and therefore, the quality of the mating surfaces and the thickness of the layer of thermally conductive paste are critical.

The thickness of paste at the point of contact shall not exceed the value of roughness on the mating components; and the paste should be applied with even layer to the degreased surface and smeared for guaranteed filling of all surface irregularities.

One of the main advantages of the presented method to reduce thermal stress in the cutting zone through removing heat from the cutting insert to the tool holder is its low cost and no need to use complicated equipment. In the process of the tool operation, the thermally conductive interface does not require frequent replacement, since it is sufficient to form it for the whole lifetime of cutting insert, and it should be replaced only with the dismantling and replacement of worn-out cutting insert. However, prior to the formation of a new layer of thermally conductive interface, the old paste should be removed with a detergent, and the surfaces should be completely degreased and dried. The above shows the great technological effectiveness of use of thermally conductive paste in production environment [2]. The studies have shown that the use of thermally conductive paste in composite cutter improves heat transfer from the cutting zone and reduces the thermal stress of tool cutting wedge. This greatly increases the efficiency of the cutting tool during wheel turning.

5. Deposition of nanoscale composite multilayer coatings (NMCC)

Nanoscale composite multilayer coatings were deposited on carbide inserts using filtered cathodic vacuum arc deposition (FCVAD) with the vacuum-arc unit VIT-2 [9, 10, 12, 13, 19]. The study used a three-component nanoscale composite multilayer coatings (NMCC) system, comprising outer (wear-resistant) layer, intermediate layer, and adhesive layer. The developed three-component NMCCs meet at best the dual nature of coatings as an intermediate process medium between the tool material and the material being machined. The coating should at the same time increase the physical and mechanical properties of the cutting tool (hardness, heat resistance, wear resistance) and reduce thermal and mechanical effect on the contact pads, resulting in their wear. The analysis of the influence of the synthesis process parameters on various properties of composite coatings (e.g., Ti-TiN-TiCrAlN) has shown that the most important parameters are as follows: current of titanium cathode arc $I_{Ti.}$, nitrogen pressure in vacuum chamber p_N, and bias potential on the substrate (tool) during condensation of wear-resistant layer U_k. These parameters were taken as major ones for the deposition of NMCCs.

The investigation into the microstructure of NMCCs was carried out on a Jeol electron scanning microscope JSM-6480LV. The macroscopic properties of NMCCs, such as thickness, hardness, friction coefficient, and strength of coating adhesion to substrate, were determined by standard methods.

Using a portable computer tomography UPUC-2000, the temperature gradient of the developed tool system was obtained as shown in **Figure 10**. Here, a reduction in the intensity of the heat source in the NMCC can be seen with a better heat dissipation through the thermal pad.

The certification (industrial confirmation) tests of the developed tool system were carried out in turning of running surfaces of wheel sets. The tests were conducted with carbide inserts (14% TiC, 8% Co, 78% WC) without coatings, and with inserts coated with the developed Ti-TiN-TiCrAlN NMCCs. Tests were performed on the heavy machines of Rafamet UCB-125 bUBB112, the criterion of insert failure was flank wear $VB_{max} = 0.5$ mm.

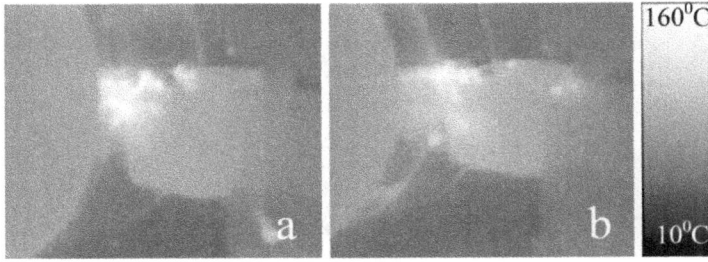

Figure 10.
A general view of heat distribution in the cutting zone at cutting speed v_c = 40 m/min, a_p = 3 mm, f = 1.0 mm/ rev [19]; a—commercial carbide inserts with multilayer coatings of modern generation and b—newly developed tool system.

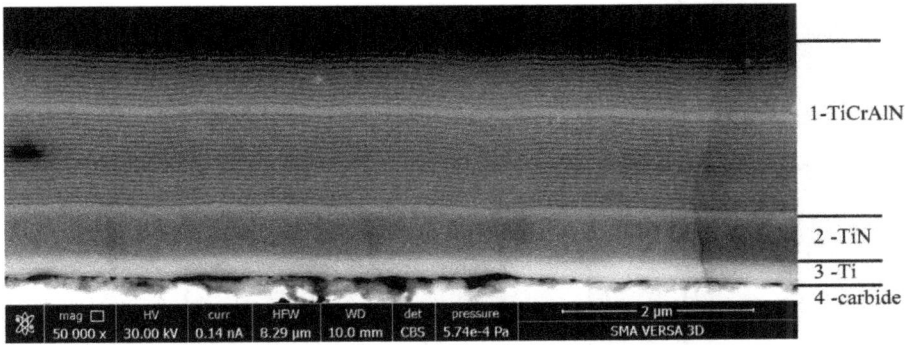

Figure 11.
SEM section image of (WC-TiC-Co) carbide insert with NMCC based on Ti-TiN-TiCrAlN [19]: 1—TiCrAlN wear-resistant layer, 2—TiN intermediate (transition) layer, 3—Ti adhesive sublayer, and 4—carbide substrate (14% TiC, 8% Co, and 78% WC).

The properties of NMCCs base on the Ti-TiN-TiCrAlN system are illustrated in **Figure 11** and in **Table 1**.

The Ti-TiN-TiCrAlN multilayer composite coating was deposited with the following process parameters: I_{Ti} = 104 A, p_N = 0.24 Pa, U_C = 42 V, and deposition time = 45 min. The analysis of the data presented in **Table 1** shows the following.

Wear-resistant layer of TiCrAlN of the tested NMCC based on the Ti-TiN-TiCrAlN system has a super multilayer architecture with sublayer thicknesses of about 15–25 nm, a columnar grain structure oriented perpendicular to the plane of TiN underlayer, in which grain sizes do not exceed 5–15 nm. The thicknesses of the sublayers of the intermediate TiN layer were also about 15–25 nm, and the sizes of its grains, as well as of grains of adhesion sublayers, did not exceed 5–15 nm. The results obtained allow classifying the multilayer composite coating of Ti-TiN-TiCrAlN as a nanocoating. The use of a vacuum arc system with filtration of vapor-ion flow FCVAD provided a significant increase in the quality of the surface of NMCC and almost a complete absence of droplets (which are dangerous defects) on the surface of the coating. This study revealed a high efficiency of the developed tool system based on double-sided (WC-TiC-Co) carbide inserts (see **Figure 8a**), with dense contact with tool holder, provided by elastic pads of reinforced ceramic-polymer material with high thermal conductivity. Tool life and coefficient of tool life variation for the developed tool system were compared with commercial tool equipped with carbide inserts with multilayer coating of the modern generation. The tool life coefficient TTL (**Figure 12**) was determined as the ratio of tool life

Material	Phase composition	Grain size, [nm]	Thickness of coating h_c, layers h_L, sublayers h_{sl} [nm]	HV, [GPa]	$F_m^{\cdot}$, [N]	ΔP, [mg/cm^2]
NMCC	Ti-TiN-TiCrAlN	5–15	$h_C = 4000$	25.0–35.0	120–130	14.7
Layer (1)	$Ti_{0.45}Cr_{0.35}Al_{0.2}N$	5–15	$h_l = 2000$; $h_{sl} = 15$–25	35.0–40.0	—	—
Layer (2)	TiN	5–15	$h_L = 1500$ $h_{sl}, = 15$–25	23.0–30.0	—	—
Layer (3)	Ti	5–15	$h_l = 500$	—	—	—

Table 1.
Test results of NMCC parameters (on the example of Ti-TiN-TiCrAlN system).

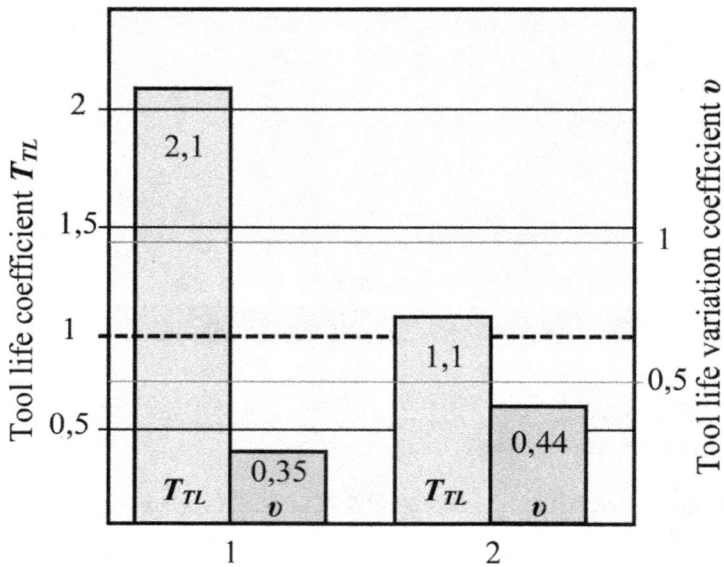

Figure 12.
Results of comparison of tool life coefficient TTL and tool life variation υ of the developed tool system with commercial carbide inserts with coatings in rough turning of wheel sets [19]; process parameters: $v_c = 50$ m/min, $f = 1.2$ mm/rev, and $a_p = 6.0$ mm. 1—developed tool system and 2—commercial carbide inserts with modern multilayer coatings.

of coated insert to tool life of an uncoated insert; and the tool life variation υ was determined as the ratio of standard deviation of tool life to its arithmetic mean value. The study showed that the developed tool system based on inserts of carbide (WC-TiC-Co) with Ti-TiN-TiCrAlN NMCCs outperformed the commercial version of carbide insert with coating of the modern generation during hard reconstruction turning of running surfaces of wheel sets (**Figure 12**). In particular, the study has shown not only the higher average tool life value (88.1 min) and tool life coefficient TTL (2.1), but also the decrease in the coefficient of tool life variation (υ = 0.355). The latter indicates the significant increase in the reliability of the developed tool system for rough turning of wheel sets.

The results of production tests of carbide cutting tools with the use of thermal pads made of NOMAKON KPTD-2 are shown in **Table 2**. The analysis of the results of the production tests also shows the increase in tool life in the developed tool system.

Part to machine	Machining type	Insert shape	Cutting parameters			Tool life improvement (%)
			a_p, mm	f, mm/rev	v_c, m/min	
New shaft of freight cars	A	SNMG 250724	5–15	0.9–1.1	80–100	20
Wheel sets with new wheel bands	B	LNMX 301940	6–8	1.1–1.3	40–50	16

A—turning-and-contouring machining of surface of neck, wheel seat and middle part of axes and B—turning-and-contouring machining along outer diameter.

Table 2.
Results of industrial tests of 14% TiC; 8% Co; 78% WC carbide inserts with NMCC and thermal pad.

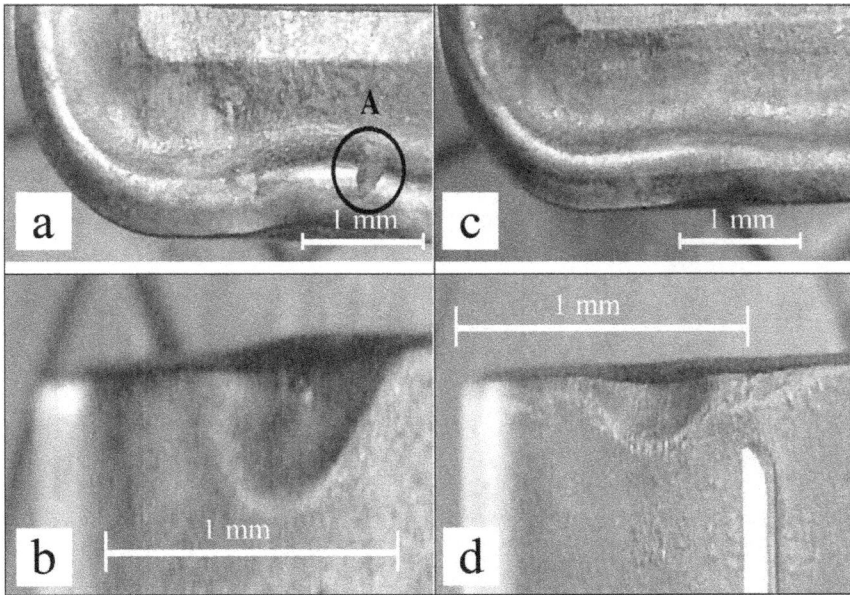

Figure 13.
Wear of carbide insert LNUX 301940 [19]: (a, c) rake, (b, d) flank, (a, b) without thermal pad, intensive plastic deformation of cutting edge, leading to cracking and brittle fracture—(zone A). (c, d) with thermal pad, less pronounced plastic deformation, mainly flank wear [2].

The effect of applied thermal pad on the nature of the wear of rake and flank faces is illustrated in **Figure 13**.

Additional tests were done with inserts AT15S carbide type LNUX 301940. These carbides were composed of 86.5% WC, 2.5% TiC, 3.6% TaC, 1.5% NbC, 5.5% Co. These tests showed a significant effect of the applied thermal pad, and the NMCC in turning of wheel sets. This is depicted in **Figure 14**.

1—uncoated inserts without thermal pad, 2—uncoated inserts with thermal pad, 3—inserts with TiN (PVD) coating, without thermal pad, 4—inserts with TiN (PVD) coating, with thermal pad, 5—inserts with Ti-TiN-TiCrAlN NMCC, without thermal pad, and 6—inserts with Ti-TiN-TiCrAlN NMCC, with thermal pad.

The studies were carried out under turning of running surface of wheel pair with v_c = 50 m/min, f = 1.2 mm/rev, and a_p = 6.0 mm (presented in **Figure 15**).

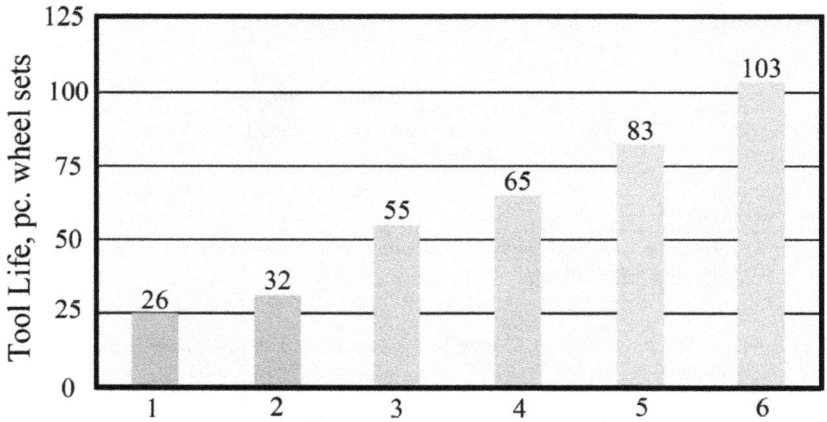

Figure 14.
Average tool life of one insert with eight cutting edges for carbide inserts type LNUX 301940, carbide AT15S [19] (process parameters: v = 50 m/min, f and = 1.2 mm/rev, and ap = 6.0 mm) [2].

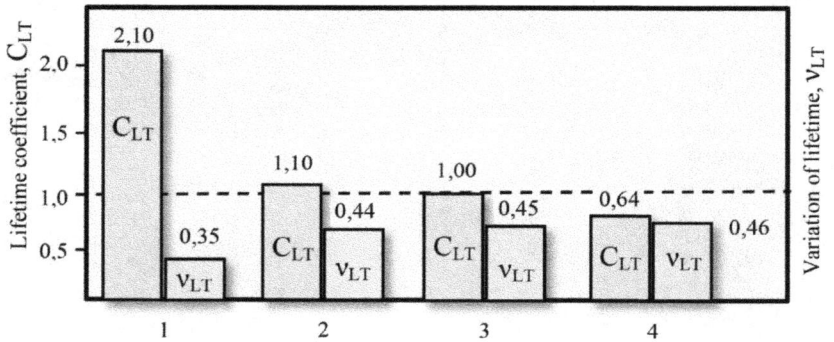

Figure 15.
Results of comparative tool life tests CLT and variations of tool life v_{LT} of the carbide inserts with standard coating of leading manufacturers (2–4) and the inserts made of carbide WC-TiC-Co with high thermal conductivity and developed NMCC (1) under rough turning of railway wheel sets [19]. 1—inserts with high thermal conductivity and developed Ti-TiN-TiAlN (technology FCVAD), 2—inserts with standard multilayered composite coating TiN-TiCN-TiN (CVD, manufacture 2), 3—inserts with standard coating TiCN-Al$_2$O$_3$-TiN (HT-CVD, manufacture 3), and 4—inserts with standard coating TiC-TiCN-TiN (HT-CVD, manufacture 4).

Evaluation of the working efficiency of cutting inserts was performed by the coefficient of wear resistance relative to the reference inserts of WC-TiC-Co without coating, in which wear resistance was taken as a unit under tests with the specified machining modes with limited flank wear land HV = 0.5 mm. The comparison was made with carbide inserts of best manufacturers with standard coatings and inserts with generated interface improving the thermal conductivity of the cemented carbide and developed NMCC.

The analysis of the studies presented in **Figure 15** allows noting the following. The analysis proved high efficiency of carbide cutting inserts in the form LNMX made of carbide grade WC-TiC-Co with thermally conductive interface and developed NMCC on the basis of a three-layer system of Ti-TiN-(Ti, Al)N in comparison with standard analogues under severe reductive turning of running surface of wheel pairs. In particular, the analysis notes not only the higher average value of lifetime of carbide tool (88.1 min) and the lifetime coefficient CLT(2.10), but also reduction of the factor of a variation of lifetime (υ_{LT} = 0.35). This indicates the significant increase in working efficiency and reliability of the tool equipped with tangential carbide inserts LNMX made of WC-TiC-Co with enhanced thermal conductivity of carbide with elaborated NMCC on the basis of the system Ti-TiN-TiAlN developed for reductive turning (roughing) of hardened (hard-drawn) surface of wheel pairs [2].

6. Conclusion

Reprofiling of the wheel tread profile is an important and large-scale manufacturing challenge. Both in milling and turning, the cutting tool is a key element ensuring efficiency and reliability of manufacturing process. The improvement of the metal cutting tools can be carried out in several directions at once, as follows:

- replacement of traditional grades of WC-TiC-Co carbides with more efficient ones based on WC-TiC-TaC-Co,

- application of special thermally conductive pads, gaskets and pastes to improve the distribution of heat flows in the cutting zone, and

- application of modern nanoscale composite multilayer coatings (NMCC).

Even higher performance can be obtained by combining the above three methods, in particular, by combining application of special thermal pads and NMCC.

The analysis of the results of laboratory and industrial tests has shown that the use of the developed tool system, including carbide inserts with NMCCs and set of structures for mounting the insert on the tool holder, including high heat-conducting ceramic-polymer pads with high thermal conductivity increased the actual contact bearing surface between the carbide insert and the holder intensifies the effective heat transfer along bearing surface of the insert. The combined effect of the increase in heat transfer and reduction of frictional heat sources due to application of the developed NMCCs showed a significant reduction of the thermal stress of the cutting system during roughing of rolling stock products. This new approach has positively transformed the character of the tool wear, and brought in an improvement of the tool life up to four times by increasing the reliability of the tool due to reduction of the coefficient of tool life variation.

The technology of modifying treatment of carbide tools was developed, which increases wear resistance of tools to high-temperature creep and tool failure because of ductile fracture of its cutting part. That is achieved by introducing between the supporting surfaces of tool holder and cutting carbide inserts a special paste (AlSink-3) on the basis of a mixture of oxides of aluminum and zinc with an silicon organic solvent that improves thermal conductivity of cemented carbide by filling air gaps between the supporting surfaces and by application of nanodispersed multilayered composite coatings, which reduce contact stresses and power of friction heat sources under severe reprofiling machining of hardened running surface of rail wheel pairs. It was found out that wear resistance of tools made of carbide grade (WC-TiC-Co) with thermally conductive interface and nanodispersed multilayered composite coating on the basis of Ti-TiN-TiAlN was up to two times higher than wear resistance of the inserts with standard coatings.

Acknowledgements

This research was financed by the Ministry of Education and Science of the Russian Federation (Leading researchers, project 16.9575.2017/6.7).

Conflict of interest

The authors declare no conflict of interest.

Author details

Alexey Vereschaka[1]*, Popov Alexey[2], Grigoriev Sergey[1], Kulikov Mikhail[2] and Sotova Catherine[1]

1 Moscow State Technological University (STANKIN), Moscow, Russia

2 Moscow State University of Railway Engineering (MIIT), Moscow, Russia

*Address all correspondence to: dr.a.veres@yandex.ru

IntechOpen

References

[1] Zakharov BV, Popov AY. Research work to improve the reliability and durability of mills on the machine tool KZh-20. Industrial Comparative Durability Tests of Carbide Inserts of Milling Cutters for Turning Wheelsets on Machines of Type KZh-20: Report on Research with Recommendations/MIIT; 123/96. Moscow; 1996. p. 95

[2] Vereschaka AS et al. Improvement of working capacity of carbide tools for machining rail wheel pairs. Key Engineering Materials. 2014;**581**:9-13

[3] Vereschaka AA et al. Nano-scale multilayered composite coatings for cutting tools operating under heavy cutting conditions. Procedia CIRP. 2014;**14**:239-244

[4] Adaskin AM et al. Cemented carbides for machining of heat-resistant materials. Advanced Materials Research. 2013;**628**:37-42

[5] Vereschaka AA. Improvement of working efficiency of cutting tools by modifying its surface properties by application of wear-resistant complexes. Advanced Materials Research. 2013;**712-715**:347-351

[6] Faga MG et al. AlSiTiNnanocomposite coatings developed via Arc Cathodic PVD: Evaluation of wear resistance via tribological analysis and high speed machining operations. Wear. 2007;**263**:1306-1314

[7] Zhang ZG et al. Microstructures and tribological properties of CrN/ZrN nanoscale multilayer coatings. Applied Surface Science. 2009;**255**:4020

[8] Vereschaka AA, Vereschaka AS, Anikeev AI. Carbide tools with nano-dispersed coating for high-performance cutting of hard-to-cut materials. Advanced Materials Research. 2014;**871**:164-170

[9] Beake BD, Fox-Rabinovich GS. Progress in high temperature nanomechanical testing of coatings for optimizing their performance in high speed machining. Surface & Coatings Technology. 2014;**255**:02-111

[10] Bouzakis K-D et al. Cutting with coated tools: Coating technologies, characterization methods and performance optimization. CIRP Annals - Manufacturing Technology. 2012;**61**:703-723

[11] Vereshchaka AA et al. Nano-scale multilayered-composite coatings for the cutting tools. International Journal of Advanced Manufacturing Technology. 2014;**72-1**:303-317

[12] Volkhonskii AO et al. Filtered cathodic vacuum arc deposition of nano-layered composite coatings for machining hard-to-cut materials. International Journal of Advanced Manufacturing Technology. 2015;**81**(1-4):1-15

[13] Vereschaka AS et al. Control of structure and properties of nanostructured multilayer composite coatings applied to cutting tools as a way to improve efficiency of technological cutting operation. Journal of Nano Research. 2016;**37**:51-57

[14] Fox-Rabinovich GS et al. Tribological adaptability of TiAlCrN PVD coatings under high performance dry machining conditions. Surface and Coatings Technology. 2005;**200**:1804

[15] Tabakov VP. The influence of machining condition forming multilayer coatings for cutting tools. Key Engineering Materials. 2012;**496**:80-85

[16] Vereschaka AS et al. Study of
properties of nanostructured multilayer
composite coatings of Ti-TiN-(TiCrAl)
N and Zr-ZrN-(ZrNbCrAl)N. Journal of
Nano Research. 2016;**40**:90-98

[17] Boothroyd G, Knight
WA. Fundamentals of Machining and
Machine Tools. Boca Raton: CRC Press;
2006

[18] Shaw MC. Metal Cutting Principles.
Oxford: Clarendon Press; 1989

[19] Vereschaka AA, Vereschaka AS,
Popov AY, Batako AD, Sitnikov NN,
Tverdokhlebov AS. Development of tool
system for high-performance machining
of products of railway rolling stock.
Procedia CIRP. 2016;**46**:360-363